D1268274

Design and Application of Security/Fire-Alarm Systems

JOHN E. TRAISTER

Design and Application of Security/Fire-Alarm Systems

Revised Edition

McGraw-Hill Publishing Company

New York St. Louis San Francisco Auckland Bogotá
Caracas Hamburg Lisbon London Madrid Mexico
Milan Montreal New Delhi Oklahoma City
Paris San Juan São Paulo Singapore
Sydney Tokyo Toronto

Library of Congress Cataloging in Publication Data

Traister, John E.

Design and application of security/fire-alarm systems / John E. Traister. — Rev. ed.

p. cm.

1. Fire alarms. 2. Security systems. I. Title.

TH9271.T7 1990 621.389′28—dc20

ISBN 0-07-065184-1 90-30239

1234567890 DOC/DOC 9543210

ISBN 0-07-065184-1

The sponsoring editor for this edition was Harold B. Crawford and the production supervisor was Suzanne W. Babeuf. It was set in Baskerville by University Graphics, Inc.

Printed and bound by R. R. Donnelley & Sons Company.

CONTENTS

PREFACE

Signaling techniques are not new. Methods were devised more than 5000 years ago to signal individuals and tribes of danger and of oncoming strangers; Indians used smoke signals for communicating with each other; other tribes used drums, animal horns, and other natural objects of sight or sound; bells were used extensively during the early settling of this country to announce meetings and warn of fires and other dangers. And, of course, military troops have, for years, used flags and horns (bugles) to communicate and to signal orders, like "charge," "retreat," and "assemble."

When electricity was put to practical use around the latter part of the last century, methods were devised to use electrical buzzers and bells, such as doorbells, entrance detectors, and manually operated fire-alarm signals, for signaling devices. As electrical and electronic devices matured, more sophisticated devices that enabled security and fire-alarm systems to become almost completely automated came on the market. Still, such devices were usually limited to certain specialized applications, such as in banks and school buildings.

Today, due to the growing populations, the use of security/fire-alarm systems is not only advisable, it is mandated in most populated towns and cities by national codes or local ordinances. Apartment buildings and town houses, for example, must have an adequate number of smoke detectors installed to warn occupants of any fire. Certain buildings are required to have sprinkler systems installed also, and must be designed to operate in conjunction with the fire-alarm system.

The typical fire-alarm system consists of sensors placed at strategic locations throughout a building and is connected (usually by electrical conductors) to a power source. When any of these sensors detects either heat or smoke that is above the building's norm, the sensors send a signal to the central control panel. Electronic devices in this panel in turn send signals to other devices on the system—which may be a combination of horns, bells, or sirens—or may even dial the fire department to inform it of the fire. Sprinkler valves, designed to activate at certain predetermined temperatures, also release water to certain restricted areas where the temperature of the area warrants.

Security, intrusion, or burglar alarms work in a similar manner; that is, they also have sensors. These detect an illegal entry into a building or enclosure. They are found in many different forms, such as electrical contacts that break when a window or door is opened; and devices capable of detecting motion, body heat, sounds, or vibrations. When any of these sensors is activated, it sends a signal; perhaps to ring a bell or siren to warn of an illegal entry, to dial the local police department, or both. Other exotic devices, such as tear gas and gas that induces sickness, have been incorporated into security systems to ward off intruders. Closed-circuit television cameras have also become an important part of security systems to identify the person or persons involved in the violation.

This book was originally written to acquaint the reader with the various systems available and to tell the reader how to use each type in various applications; what types are best in certain locations; and, finally, how to install them in the most proficient manner. Since the first edition was published in 1981, this book has found use in architectural and consulting engineering firms, in electrical contracting firms, and in other firms that install security systems. Furthermore, electrical/electronic technicians—both students and workers—have found the book to be a handy reference. Manufacturers and their sales representatives usually keep a copy close at hand. In fact, any property owner—anyone who has anything he or she wants to protect—will find this book invaluable.

Readers will learn how to select and design a security/fire-alarm system for practically any application. Once the system is designed, chapters in this book show how to estimate the installation charge. Then, once the system is installed, troubleshooting techniques help keep the system in operation.

The chapters are arranged in a structural sequence intended to satisfy the immediate and fundamental needs of the beginning student in the field of security/ fire-alarm systems, as well as the design engineer, architect, contractor, or installer.

A deep and grateful bow is made in the direction of the numerous manufacturers who contributed many of the illustrations and much of the reference material during the preparation of this book.

John E. Traister

Design and Application of Security/Fire-Alarm Systems

1

BASIC CONSIDERATIONS

The design and installation of security and fire-alarm systems employs a wide variety of techniques, often involving special types of equipment and materials designed for specific applications. Many systems operate on low-voltage circuits but are installed similarly on conventional electrical circuits for light and power. All installations, when used in buildings, must conform to applicable National Electrical Code (NEC) requirements (especially those codes covered in Chaps. 6 and 7 of the NEC), local ordinances, and instructions provided by security and fire-alarm system manufacturers and design engineers.

CLASSIFICATION OF SIGNAL CIRCUITS

A signal circuit used for a security or fire-alarm system may be classified as *open circuit* or *closed circuit*. An open circuit is one in which current flows only when a signal is being sent. A closed-circuit system is one in which current flows continuously, except when the circuit is opened to allow a signal to be sent.

All alarm systems have three functions in common: detection, control, and annunciation (or alarm) signaling. Many systems incorporate switches or relays that operate when entry, movement, pressure, infrared-beam interruption, and other intrusions occur. The control senses the operation of the detector with a relay and produces an output that may operate a bell, siren, silent alarm such as telephone dialers to law enforcement agencies, or other signal. The controls frequently contain ON/OFF switches, test meters, time delays, power supplies, standby batteries, and terminals for connecting the system together. The control output usually provides power on alarm to operate signaling devices or switch contacts for silent alarms. See the diagram in Fig. 1-1.

An example of a basic closed-circuit alarm system is shown in Fig. 1-2. The detection, or input, subdivision in this drawing shows exit/entry door or window

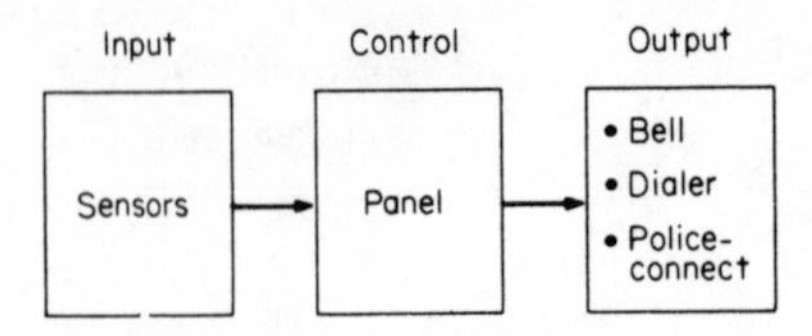

FIGURE 1-1. Basic subdivisions of an alarm system.

contacts. However, the detectors could just as well be smoke or heat detectors, switch mats, ultrasonic detectors, etc.

The control subdivision for the system in Fig. 1-2 consists of switches, relays, a power supply, a reset button, and related wiring. The power supply shown is a 6-V nickel-cadmium battery that is kept charged by a plug-in transformer unit. Terminals are provided on the battery housing to accept 12-Vac charging power from the plug-in transformer which provides 4- to 6-V power for the detection (protective) circuit and power to operate the alarm or output subdivision.

Figure 1-3 shows another closed-circuit system. The protective circuit consists of a dc energy source, any number of normally closed intrusion-detection contacts (wired in series), a sensitive relay (R_1), and interconnecting wiring. In operation, the normally closed intrustion contacts are connected to the coil of the sensitive relay. This keeps the relay energized, holding its normally closed contacts open against spring pressure—the all-clear condition of the protective circuit. The opening of any intrusion contact breaks the circuit, which de-energizes the sensitive relay and allows spring force to close the relay contacts. This action initiates the alarm.

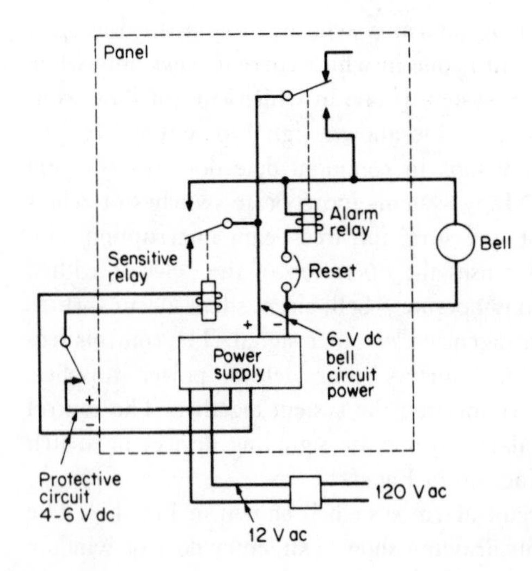

FIGURE 1-2. Basic closed-circuit security alarm system.

The key-operated switch shown in the circuit in Fig. 1-3 is provided for opening the protective circuit for test purposes. A meter (M) is activated when the switch is set to CIRCUIT TEST. The meter gives a current reading only if all intrusion contacts are closed. All three sections of the switch (S_1, S_2, S_3) make contact simultaneously as the key is turned.

Opening of intrusion contacts is not the only event that causes the alarm to activate. Any break in protective-circuit wiring or loss of output from the energy source has the same effect. The circuit is broken which de-energizes the sensitive relay and allows spring force to close the relay contacts, thus sounding the alarm. Any cross or short circuit between the positive and negative wires of the protective circuit also keeps current from reaching the relay coil and causes dropout which again sounds the alarm.

Other components of the alarm circuit in Fig. 1-3 include a second energy source, an alarm bell, and a drop relay (R_2). When the keyed switch is at ON, dropout of the sensitive relay R_1 and closing of its contacts completes a circuit to energize the coil of drop relay R_2. Closing of the drop relay's normally open contacts rings the bell and latches in the drop-relay coil so that R_2 stays energized even if the protective circuit returns to normal and opens the sensitive relay's contacts. As a result, the bell continues to ring until the key switch is turned away from ON to break the latching connection to the R_2 coil.

Drop relays often have additional contacts to control other circuits or devices. The extra contacts in the circuit in Fig. 1-3 are for turning on lights, triggering an automatic telephone dialer, etc. But the main two functions of the drop relay

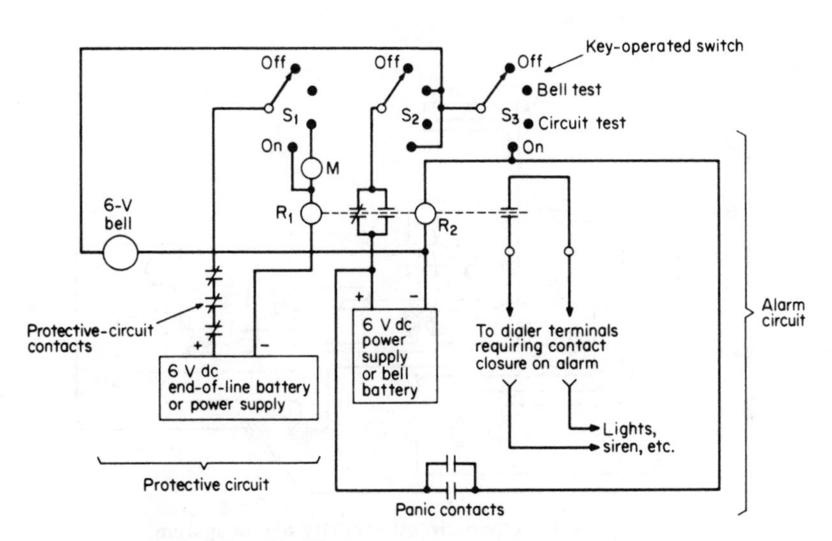

FIGURE 1-3. Closed-circuit security alarm system.

are actuation of the alarm and latching the coil to keep the circuit in the alarm condition.

Almost all burglar systems use a closed-loop protective circuit, so only a very brief description of an open-circuit system will be given here. In general, the system consists of an annunciator connected to a special design contact on each door and window and a relay so connected that when any window or door is opened it will cause current to pass through the relay. The relay, in turn, will operate to close a circuit on a bell, horn, or other type of annunciator which will continue to sound until it is shut off, thereby alerting the occupants or law enforcement agencies.

The wiring and connections for the open-circuit system are shown in Fig. 1-4. This wiring diagram shows three contacts, but any number can be added as needed. Closing any one of the contacts completes the power circuit through the winding of the proper annunciator drops, the constant-ringing switch, the constant-ringing relay, the alarm bell, and the bell-cutoff switch. The current through the winding of the constant-ringing relay operates to complete a circuit placing the alarm bell directly across the battery or other power source so the bell continues to ring until the cutoff switch is opened. At the same time, current in another set of wires operates a relay that closes an auxiliary circuit to operate other devices, such as lights and automatic telephone dialer.

A spring-type contact for open-circuit operation is shown in Fig. 1-5. This device is set in the window frame so that cam *c* projects outward. When the window is raised, the cam pivots and is pressed in and makes contact with spring *s*, which is insulated from the plate by a washer at the lower end and is held free

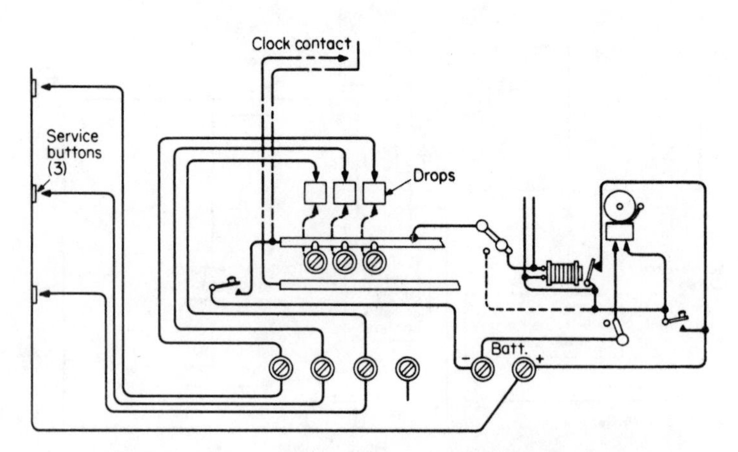

FIGURE 1-4. Open-circuit security alarm system.

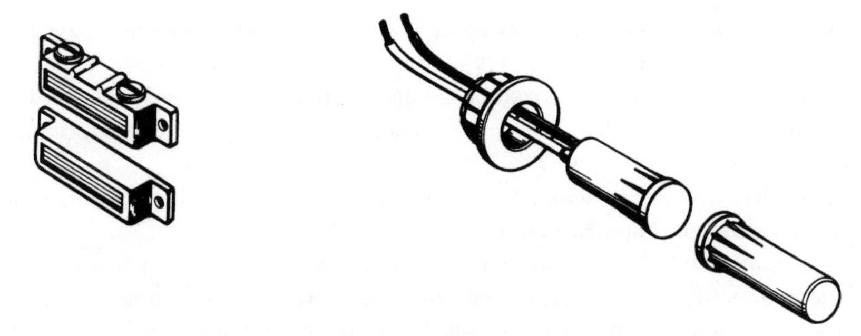

FIGURE 1-5. Spring-type contact for open-circuit operation.

from *c* by the insulating wheel *w*. This contact is connected in series with the power source and the annunciator; that is, one wire is connected to the plate and the other to the spring. An old type of door contact for an open-circuit system is shown in Fig. 1-6.

FIRE-ALARM SYSTEMS

A fire-alarm system consists of sensors, a control panel, an annunciator, and the related wiring to connect the components. Fire-alarm systems generally may be divided into four types: noncoded, master-coded, selective-coded, and dual-coded.

Each of these four types has several functional features so designed that a specific system may meet practically any need to comply with local and state fire codes, statutes, and regulations.

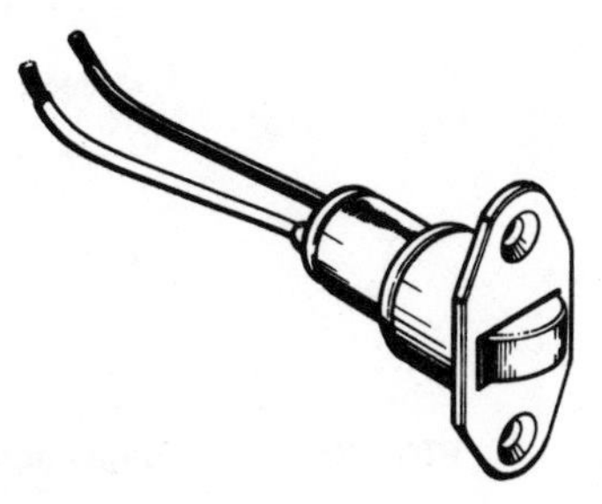

FIGURE 1-6. Old type of door contact for an open-circuit security system.

In a noncoded system, an alarm signal is sounded continuously until manually or automatically turned off.

In a master-coded system, a common-coded alarm signal is sounded for not less than three rounds. The same code is sounded regardless of the alarm-initiating device activated.

In a selective-coded system, a unique coded alarm is sounded for each firebox or fire zone on the protected premises.

In a dual-coded system, a unique coded alarm is sounded for each firebox or fire zone to notify the building's personnel of the location of the fire, while noncoded or common-coded alarm signals are sounded separately to notify other occupants to evacuate the building.

A fire-alarm system riser diagram is shown in Fig. 1-7. Basically, if any smoke detector senses smoke or if any manual striking station is operated, all bells within the building will ring, indicating a fire. At the same time, the magnetic door switches will release the smoke doors to help block smoke and/or drafts. This

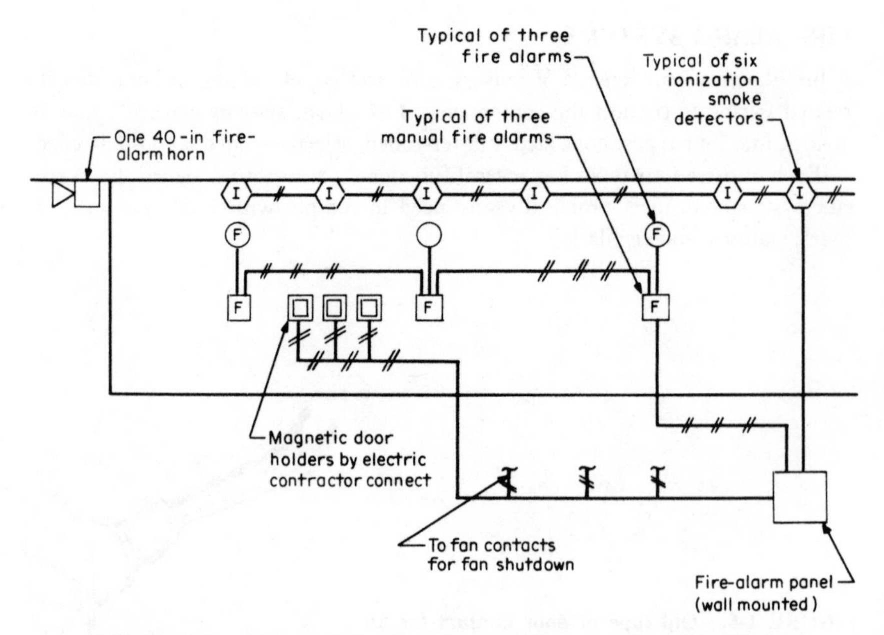

FIGURE 1-7. Diagram of a fire-alarm system riser for a courthouse.

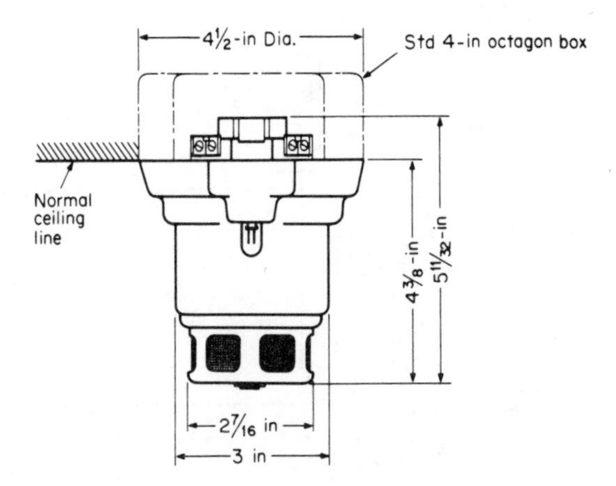

FIGURE 1-8. Ionization smoke detector.

system is also connected to a water-flow switch on the sprinkler system. If the sprinkler valves are activated causing a flow of water in the system, the fire-alarm system will again go into operation energizing all bells and closing smoke doors.

Ionization smoke detectors (Fig. 1-8) may be used in place of conventional smoke detectors or they may be used in combination with smoke detectors. The ionization smoke detectors are more sensitive than the conventional smoke detectors.

COMPONENTS OF SECURITY/FIRE-ALARM SYSTEMS

Wire sizes for the majority of low-voltage systems range from no. 22 to no. 18 AWG. However, where larger-than-normal currents are required or when the distance between the outlets is long, it may be necessary to use wire sizes larger than specified to prevent excessive voltage drop. Voltage-drop calculations should be made to determine the correct wire size for a given application—even on low-voltage circuits.

The wiring of any alarm system is installed like any other type of low-voltage system; that is, locating the outlets, furnishing a power supply, and finally interconnecting the components with the proper size and type of wire.

Most closed systems use two-wire no. 22 or no. 24 AWG conductors and are color-coded to identify them. A no. 18 pair normally is adequate for connecting bells or sirens to controls if the run is 40 ft (12 m) or less. Many, however, prefer to use no. 16 or even no. 14 cable.

A summary of the various components for a typical security/fire-alarm system is depicted in Fig. 1-9. The following list gives a description of these components.

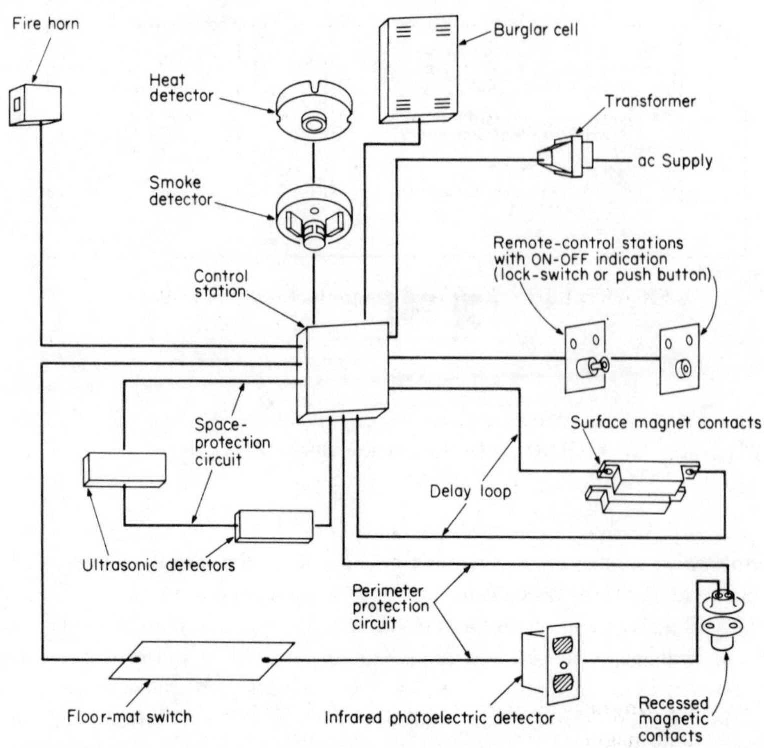

FIGURE 1-9. Various components for a typical security/fire-alarm system.

Control Station: This is the heart of any security system since it is the circuitry in these control panels that senses a broken contact and then either sounds a local bell or horn or omits the bell for a silent alarm. Most modern control panels use relay-type controls to sense the protective circuits and regulate the output for alarm-sounding devices. They also contain contacts to actuate other deterrent or reporting devices and a silent holdup alarm with dialer or police-connected reporting mechanism.

Power Supplies: Power supplies vary for different systems, but in general they consist of rechargeable 6-Vdc power supplies for burglar-alarm systems. The power packs usually contain nickel-cadmium batteries that are kept charged by 12-Vac input from a plug-in or otherwise connected transformer to a 120-V circuit. The better power supplies have the capability of operating an armed system for 48 hours or more without being charged and still have the capacity to ring an alarm bell for 30

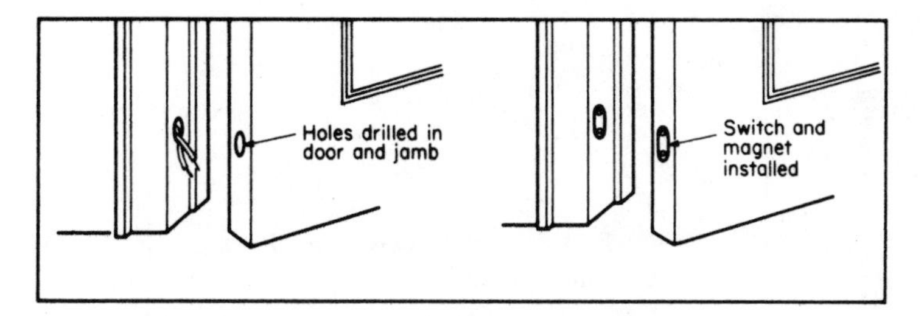

FIGURE 1-10. Recessed magnetic contacts in door.

minutes or longer. Power supplies are obviously used in conjunction with a charging source and supply power for operation of the alarm system through the control panel.

Recessed Magnetic Contacts in Door (Fig. 1-10): Holes are drilled in the door and in the casing, one directly across from the other, and a pair of wires from the positive side of the protective circuit is run out through the switch hole. The switch and magnet are then installed with no more than a ⅛-in (0.3 cm) gap between them.

Recessed Magnetic Contacts in Casement Window (Fig. 1-11): A switch and magnet are installed as in the door, preferably in the top of the window and underside of the upper window casing, where they will be least noticeable.

Surface-Mounted Magnetic Contacts on Double-Hung Window (Fig.1-12): A switch is mounted on the window casing with a magnet on the window casing and a magnet on the window. As long as the switch and magnet are parallel and in close proximity when the window is shut, they may be oriented side-to-side, top-to-side, or top-to-top.

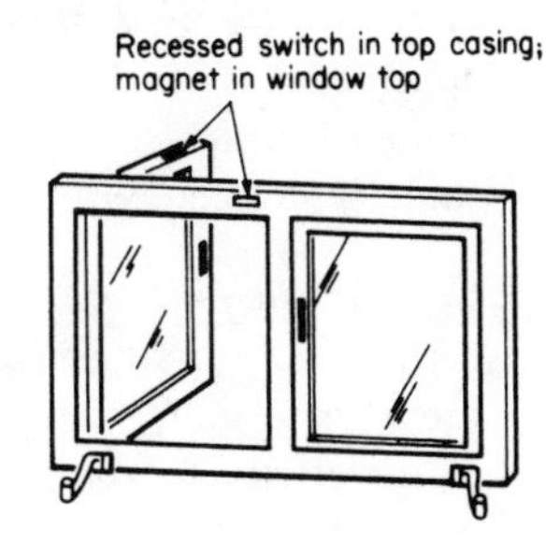

FIGURE 1-11. Recessed magnetic contacts in casement window.

FIGURE 1-12. Surface-mounted magnetic contacts on double-hung window.

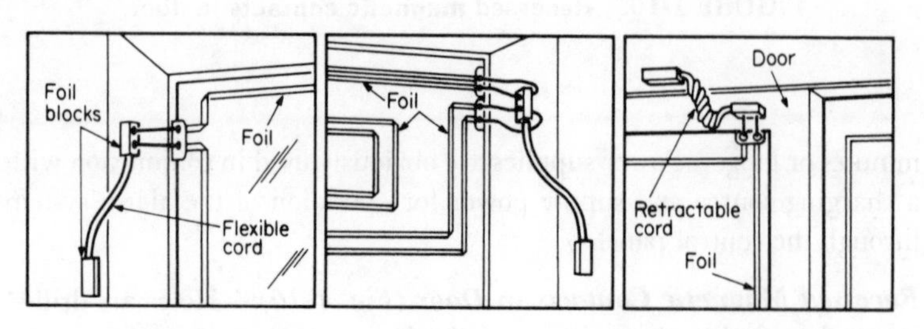

FIGURE 1-13. Conductive foil on glass doors.

Conductive Foil on Glass Doors (Fig. 1-13): A self-adhesive foil block (terminator) on the door is connected to a similar unit on the door frame by a short length of flexible cord to allow for door movement. The foil is connected in the positive conductor of the protective circuit and is adhered to the glass parallel to and about 3 in (7.6 m) from the edge of the glass, using recommended varnish. Breaking the glass breaks the foil and opens the circuit. To provide more coverage, a double circuit of foil may be taken from the foil block. Coiled, retractable cords are available for use between foil blocks to allow for sliding-door travel.

Complete Glass-Door Protection (Fig. 1-14): A glass door with a glass transom may be protected by a combination of magnetic contacts and foil.

Surface-Mounted Magnetic Contacts on Door (Fig. 1-15): Where appearance is not the most important consideration, the use of a surface-mounted switch (on the door frame) and a magnet (on the door) will simplify installation.

Conductive Foil on Picture Windows (Fig. 1-16): Where a window does not open, a single run of foil is connected to a foil block on the glass, frame, or wall. When the foil crosses over a frame member, a piece of plastic electrical tape should be used to provide an insulated crossover surface for the foil.

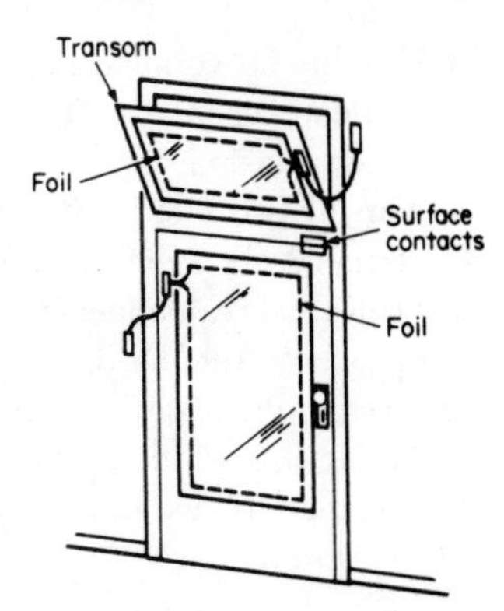

FIGURE 1-14. Complete glass-door protection.

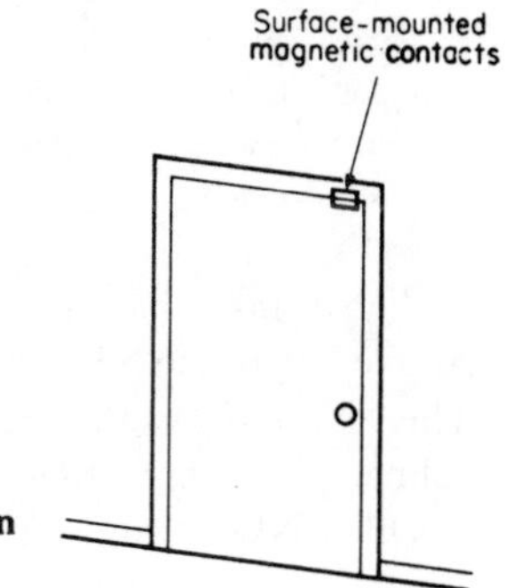

FIGURE 1-15. Surface-mounted magnetic contacts on door.

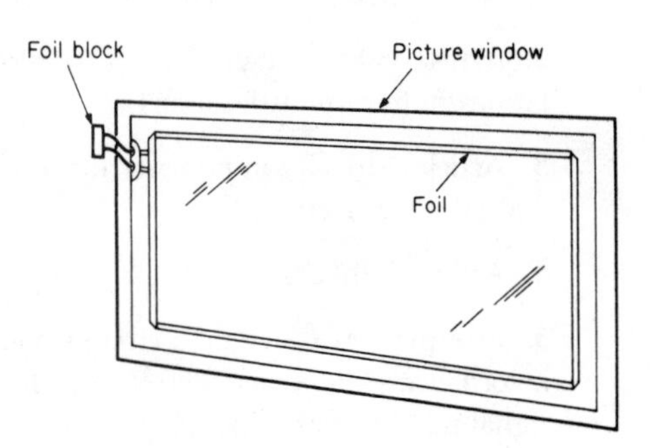

FIGURE 1-16. Conductive foil on picture windows.

These and other components are discussed in more detail in the chapters to follow.

NATIONAL ELECTRICAL CODE REQUIREMENTS

Because of the potential fire and explosion hazards caused by the improper handling and installation of electrical wiring, certain rules in the selection of materials and quality of workmanship must be followed, as well as precautions for safety. To standardize and simplify these rules and provide some reliable guide for electrical construction, the National Electrical Code (**NEC**) was developed.

The **NEC** is published (and frequently revised) by the NFPA (National Fire Protection Association), 470 Atlantic Avenue, Boston, MA 02210. It contains specific rules and regulations intended to help in the practical safeguarding of persons and property from hazards arising from the use of electricity, including low-voltage used in the majority of security/fire-alarm systems.

Article 725 of the **NEC** covers remote-control, signaling, and power-limited circuits that are not an integral part of a device or appliance. The **NEC** (Section 725-1) states:*

> The circuits described herein [Article 725] are characterized by usage and electrical power limitations which differentiate them from light and power circuits and, therefore, special consideration is given with regard to minimum wire sizes, derating factors, overcurrent protection, and conductor insulation requirements.

Those installing security/fire-alarm systems should become familiar with Article 725 of the **NEC** as well as Article 760, Fire Protective Signaling Systems. This article covers the installation of wiring and equipment of fire protective signaling systems operating at 600 volts or less.

Other **NEC** articles of interest to security/fire-alarm installers include:

1. Section 300-21 (Spread of Fire or Products of Combustion).

2. Articles 500 through 516 and Article 517, Part G (dealing with installations in hazardous locations).

3. Article 110 (Requirements for Electrical Installations) and Article 300 (Wiring Methods).

4. Article 310 (Conductors for General Wiring).

5. Fire-protective signaling circuits and equipment shall be grounded in accordance with Article 250, except for dc-power limited fire protective signaling circuits having a maximum current of 0.03 amperes.*

6. The power supply of nonpower-limited fire-protection signaling circuits shall comply with Chapters 1 through 4 and the output voltage shall not be more than 600 volts, nominal.*

7. Conductors of no. 18 and no. 16 size shall be permitted to be used provided they supply loads that do not exceed the ampacities given in

*Reprinted by permission from NFPA 70-1981, National Electrical Code®, Copyright © 1980, National Fire Protection Association, Boston, Mass.

Table 402-5 and are installed in a raceway or a cable approved for the purpose. Conductors larger than no. 16 shall not supply loads greater than the ampacities given in Tables 310-16 through 310-19.

8. When only nonpower-limited fire-protective signaling circuits and Class 1 circuits are in a raceway, the number of conductors shall be determined in accordance with Section 300-17. The derating factors given in Note 9 to Tables 310-16 through 310-19 shall apply if such conductors carry continuous loads.*

9. Where power-supply conductors and fire-protective signaling circuit conductors are permitted in a raceway in accordance with Section 760-15, the number of conductors shall be determined in accordance with Section 300-17. The derating factors given in Note 8 to Tables 310-16 through 310-19 shall apply as follows:

a. To all conductors when the fire-protective signaling circuit conductors carry continuous loads and the total number of conductors is more than three.

b. To the power-supply conductors only when the fire-protective signaling circuit conductors do not carry continuous loads and the number of power-supply conductors is more than three.

10. Where fire-protective signaling circuit conductors are installed in cable trays, comply with Sections 318-8 through 318-10.

2

CONTROLS

Devices used to control security/fire-alarm systems vary from simple toggle switches to complex systems utilizing components such as relays, timers, magnetic contacts, and so forth. This chapter is designed to acquaint the reader with some of the control circuits normally found in security and fire-alarm systems.

FIRE-ALARM SYSTEMS

Fire-alarm systems for commercial and industrial use usually fall into four basic categories: noncoded, master-coded, selective-coded, and dual-coded. Each of these four systems has several functional features to accommodate a building's needs and to satisfy local and state fire codes, statutes, and regulations.

In a noncoded system, an alarm signal is sounded continuously until manually or automatically turned off.

In a master-coded system, a common-coded alarm signal is sounded for not less than three rounds. The same code is sounded regardless of the alarm-initiating device activated.

In a selective-coded system, a unique coded alarm is sounded for each firebox or fire zone on the protected premises.

In a dual-coded system, a unique coded alarm is sounded for each firebox or fire zone to notify the owner's personnel of the location of the fire, while noncoded or common-coded alarm signals are sounded separately to notify other occupants to evacuate the building.

One of the best ways to understand the control function of the various fire-alarm systems is to analyze existing systems. For example, Fig. 2-1 shows a partial floor plan of a nursing home. This portion of the floor plan shows the fire-alarm panel, smoke detectors (designated SD), striking stations, gongs (bells), and mag-

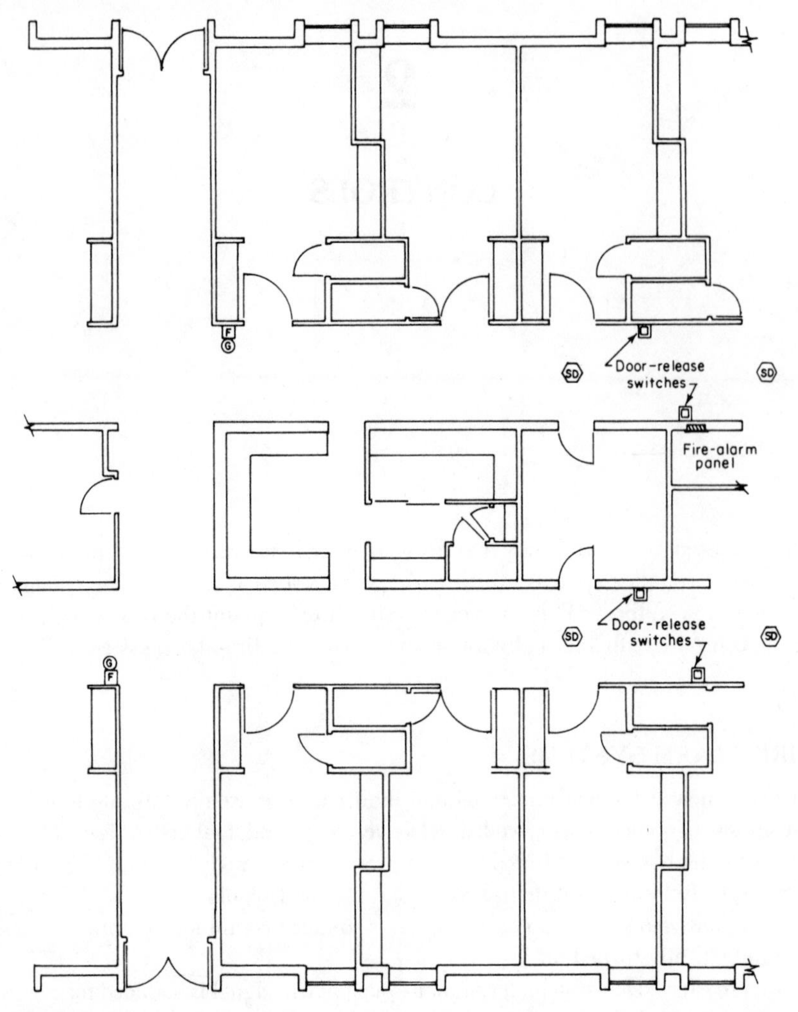

FIGURE 2-1. Partial floor plan of a nursing home showing the fire-alarm system.

netic door-release switches. The fire-alarm riser diagram in Fig. 2-2 shows all devices connected to this system, along with the wiring for each.

Basically, if any smoke detector senses smoke or if any manual striking station is operated, all bells within the building sound, indicating a fire. At the same time, the magnetic door switches release the smoke doors to help block smoke and/or drafts. This system is also connected to a water-flow switch on the sprinkler system. If the sprinkler valves are activated causing a flow of water in the system, the

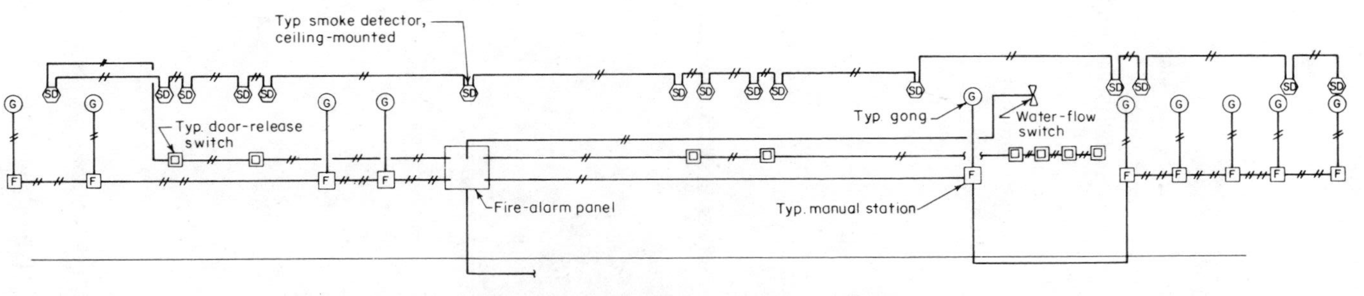

FIGURE 2-2. Fire-alarm riser diagram for the system shown in Fig. 2-1.

fire-alarm system will again go into operation energizing all bells and closing all smoke doors.

CONTROL PANELS

A typical burglar/fire-alarm panel is shown in Fig. 2-3. This particular panel is designed for combined burglar-, fire-, and panic-alarm systems. This panel, and most others, operate on low-voltage alternating current from a plug-in transformer. Many systems also have a rechargeable or dry cell battery pack for backup power should the ac source fail.

A wiring diagram for a rechargeable 6-Vdc power supply is shown in Fig. 2-4. Note that the transformer is plugged into a 120-Vac outlet which provides 12Vac on its secondary side. One terminal from the transformer connects to the charging circuit (1) while the other lead connects to one side of the battery (B_1). Fuse F_1 and resistor R_2 offer 6-A circuit protection in this case. R_2 also provides short-circuit protection.

The protective-circuit contacts from terminal 5 utilize a 100-Ω, 2-W resistor

FIGURE 2-3. Typical burglar/fire-alarm panel.

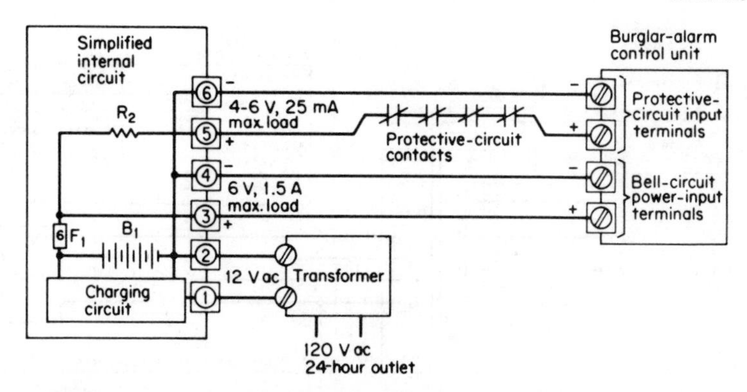

FIGURE 2-4. Wiring diagram for a rechargeable 6-Vdc power supply.

in the positive feed to each circuit to keep a cross on any one zone from affecting other zones. A detail of this connection is shown in Fig. 2-5.

Obviously, the heart of a fire-alarm system is the master control panel. To this panel are connected various detector and alarm circuits, as shown in Fig. 2-6. In this case, the primary power is taken from an unswitched three-wire 120/240 Vac distribution line. The initiating and alarm circuits are connected to the neutral

FIGURE 2-5. There are 100-Ω resistors connected in the positive feed to each circuit to keep a cross on any one zone from affecting other zones.

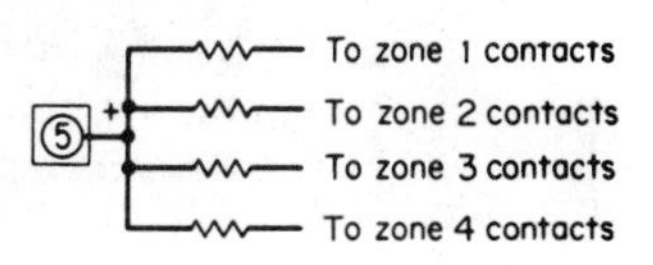

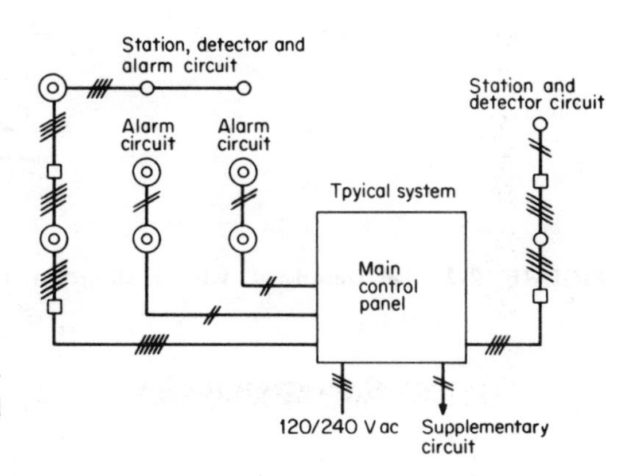

FIGURE 2-6. Wiring diagram of a master control panel.

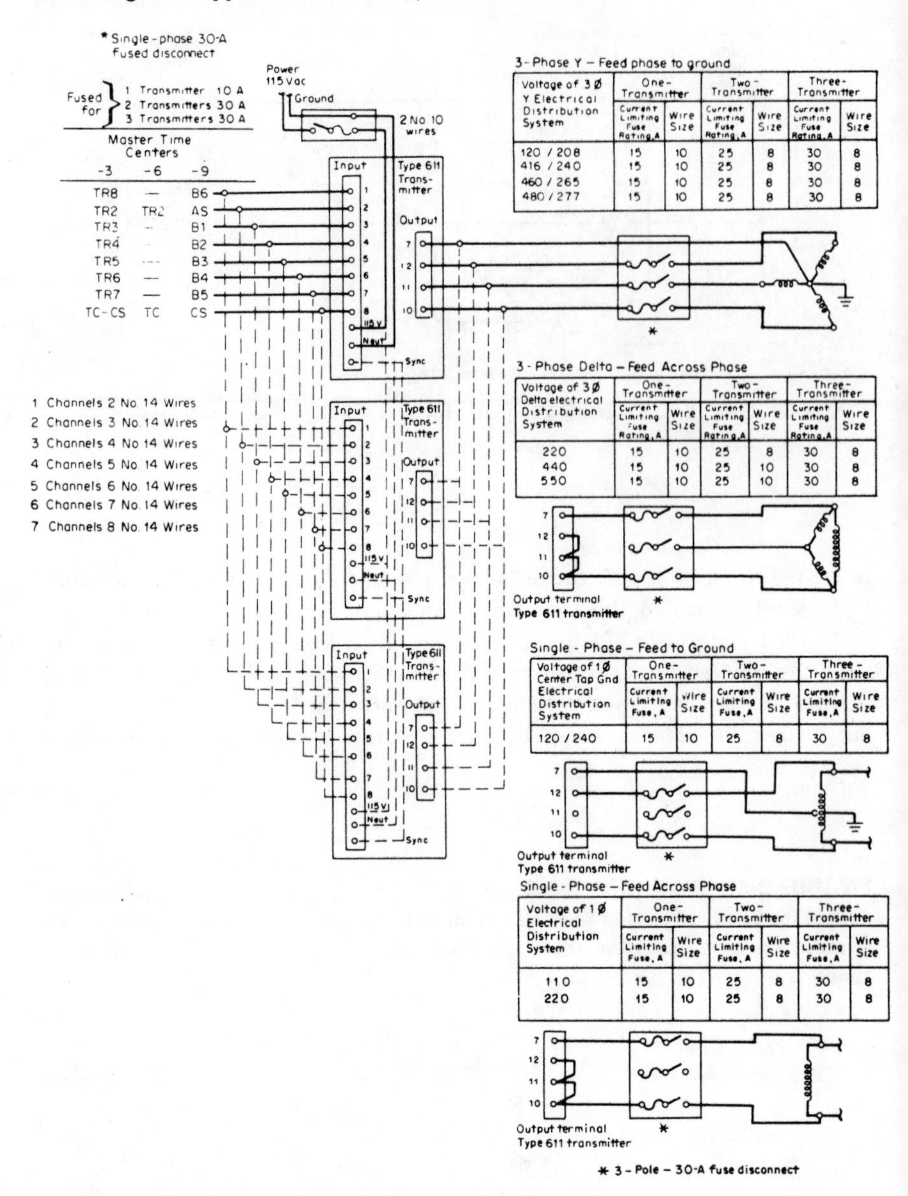

3-Phase Y – Feed phase to ground

Voltage of 3Ø Y Electrical Distribution System	One-Transmitter		Two-Transmitter		Three-Transmitter	
	Current Limiting Fuse Rating, A	Wire Size	Current Limiting Fuse Rating, A	Wire Size	Current Limiting Fuse Rating, A	Wire Size
120 / 208	15	10	25	8	30	8
416 / 240	15	10	25	8	30	8
460 / 265	15	10	25	8	30	8
480 / 277	15	10	25	8	30	8

3-Phase Delta – Feed Across Phase

Voltage of 3Ø Delta electrical Distribution System	One-Transmitter		Two-Transmitter		Three-Transmitter	
	Current Limiting Fuse Rating, A	Wire Size	Current Limiting Fuse Rating, A	Wire Size	Current Limiting Fuse Rating, A	Wire Size
220	15	10	25	8	30	8
440	15	10	25	10	30	8
550	15	10	25	10	30	8

Single-Phase – Feed to Ground

Voltage of 1Ø Center Tap Gnd Electrical Distribution System	One-Transmitter		Two-Transmitter		Three-Transmitter	
	Current Limiting Fuse, A	Wire Size	Current Limiting Fuse, A	Wire Size	Current Limiting Fuse, A	Wire Size
120 / 240	15	10	25	8	30	8

Single-Phase – Feed Across Phase

Voltage of 1Ø Electrical Distribution System	One-Transmitter		Two-Transmitter		Three-Transmitter	
	Current Limiting Fuse, A	Wire Size	Current Limiting Fuse, A	Wire Size	Current Limiting Fuse, A	Wire Size
110	15	10	25	8	30	8
220	15	10	25	8	30	8

FIGURE 2-7. (*a*) Detail of wiring diagram of a master control panel.

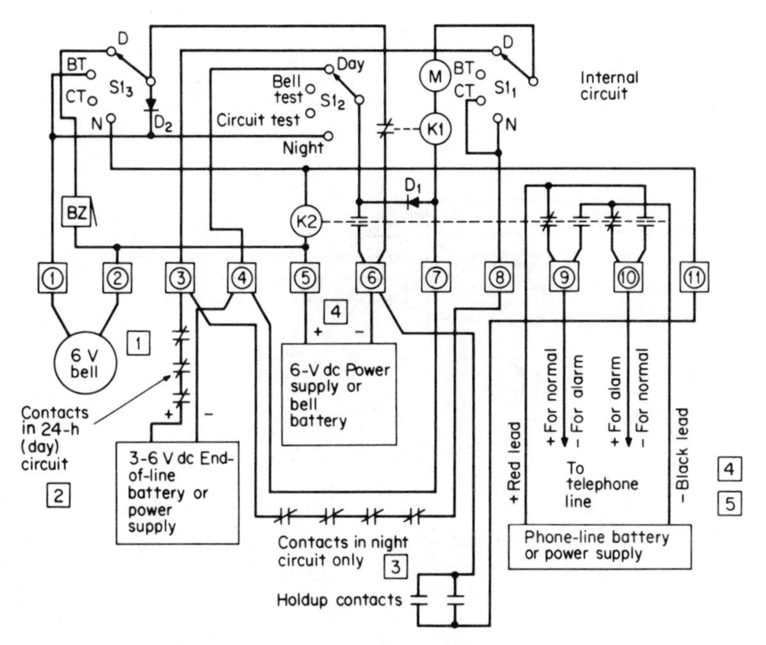

(*b*) Schematic drawing of a day-night police panel.

ground and to one leg of the main circuit. The trouble-indicator circuits are connected to the neutral ground and to the opposite leg of the circuit.

When an automatic detector or manual station is activated, the contacts close to complete a circuit path and apply 120 Vac to the alarm control circuits in the main panel. This includes a synchronous motor on some systems, which immediately operates cam assemblies that cause the alarm circuit switch contacts to make and break in a code sequence (if a code sequence is used). Additional cam-controlled switches stop the motor and alarm signals after a predetermined time lapse and actuate the alarm buzzer on the main panel.

Most control panels contain a supplementary relay control for connection to an external auxiliary circuit providing its own electrical power. The relay usually has a single-pole double-throw contact, which operates in step with the master code signal. The circuit may be used to activate other auxiliary alarms or controls, such as a city fire-department connection, fan shutdown, or door release. A schematic wiring diagram of a typical system is shown in Fig. 2-7.

A schematic drawing of a day-night police panel is shown in Fig. 2-7*b* while key switch operating sequences are depicted in Fig. 2-8. In general, any DAY circuit contact opening sounds the buzzer in the panel but does not ring the alarm bell or disturb police. A holdup contact closure sends a silent police alarm.

On the BELL TEST circuit, the bell can be rung for a test to check the power

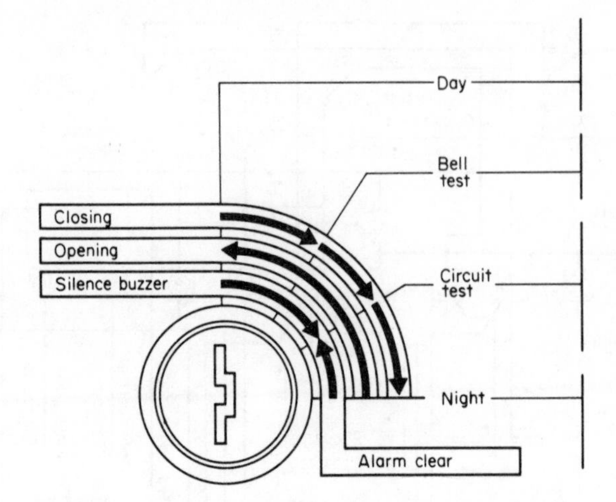

FIGURE 2-8. Key switch-operating sequences of the system in Fig. 2-7*b*.

source and wiring without disturbing the police, but the holdup circuit remains armed during this test.

During the CIRCUIT TEST sequence, the holdup circuit remains armed and the meter shows the current through DAY and NIGHT circuits combined when all contacts are closed. A reading on this particular circuit should be from 2 to 6 mA.

Any contact opening (or cross) in the DAY or NIGHT circuits rings the alarm bell and sends the police alarm. This alarm latches on until the key switch is turned back to CIRCUIT TEST or beyond.

ENTRY/EXIT DELAY MODULE

Solid-state entry/exit delay modules eliminate the need to install a shunt lock across any entry/exit door contacts in a security system. Door contacts are connected to the module, which in turn is wired into the protective circuit. Separately adjustable exit and entry delay periods allow the user to turn the system on and leave and then enter and shut the system off without causing alarms.

The module is installed in the alarm system control cabinet as shown in Fig. 2-9. It operates on current from the system's bell battery or power supply and is controlled by the switch functions available in any conventional control unit. It works like a normally closed contact in the negative side of the protective circuit, with all the protective contacts except the entry/exit door contacts wired into the positive side of the circuit. Opening of any positive-side contacts causes an instant alarm, but the module opens the negative side to cause an alarm *only* when one of the following occurs:

1. Door contacts have opened once and are still open when the exit delay expires.

2. Door contacts open after the exit delay expires when there was not an exit during the exit delay period.

3. Door contacts open after a proper exit and the system is not shut off before the entry delay expires.

ULTRASONIC MOTION DETECTORS

Ultrasonic detectors work by flooding an area with ultrasonic energy and monitoring the "sound" that returns to the detector from the covered area. In the absence of motion, the received sound is all of a single frequency. Movement of an object in the protected space shifts the frequency of some of the reflected sound, changing the output of the receiving transducer. But such frequency shifts can also be caused by certain environmental factors that must be taken into consideration at installation if false alarms are to be avoided.

PHOTOELECTRIC SYSTEMS

A photoelectric transmitter is shown in Fig. 2-10*a* and the receiver is shown in Fig. 2-10*b*. Each requires a 12-Vac input for operating power. This type of system is intended for continuous operation, regardless of whether the alarm system to which it is connected is on or off. Clicking of the receiver relay is thus a normal indication whenever the invisible infrared beam is broken or restored.

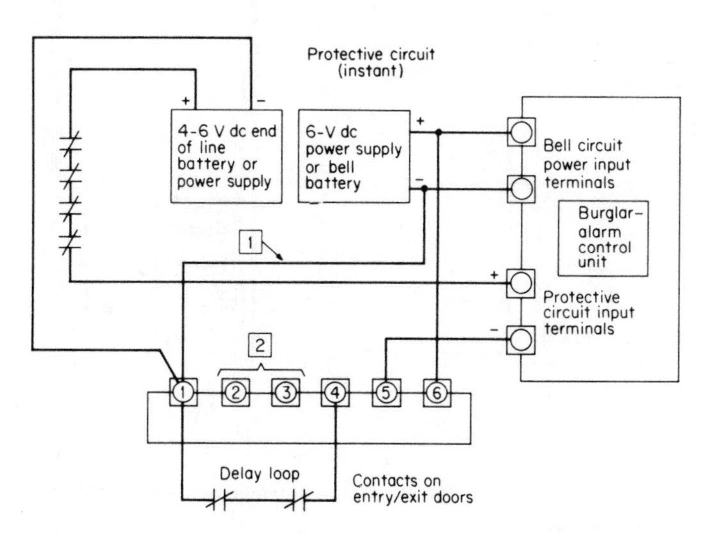

FIGURE 2-9. Connection detail of entry/exit delay module.

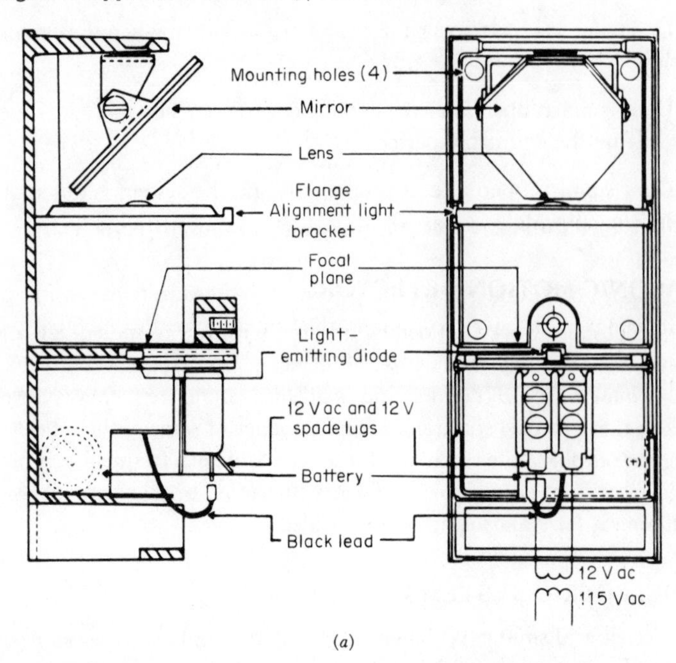

(a)

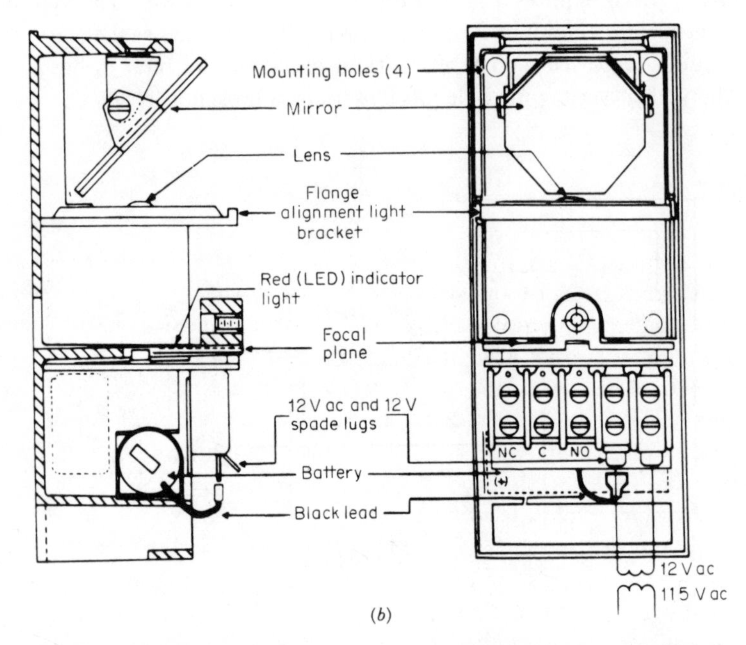

(b)

FIGURE 2-10. (a) Typical photoelectric transmitter. (b) Photoelectric receiver.

TELEPHONE DIALERS

A schematic wiring diagram of a typical telephone dialer is shown in Fig. 2-11. The dialer's two cooperating channels permit two distinct dialing and message programs. Although labeled as, and most commonly used for, separate burglar and fire alarms, the two channels can be connected and programmed for any application: medical emergency, heating-system failure, freezer warmup, water-pressure failure.

It is important to understand the priority relationship between the two channels before making trigger connections. The priority arrangement ensures transmission of the vital fire-alarm program (or other priority program on the FIRE channel) in three ways:

1. If the dialer is already operating on the BURGLAR channel when the FIRE channel is triggered, the dialer immediately switches to FIRE-channel transmission.

2. When FIRE-channel priority seizure has occurred, the dialer overrides its normal end-of-cycle stop and runs for another full cycle. This ensures transmission of the entire priority program, even if the FIRE-channel takeover occurred near the end of a BURGLAR-channel cycle.

3. Even if the dialer has stopped after transmitting the full BURGLAR-channel program and the burglar-alarm input is still present, an input on the FIRE channel causes immediate transmission of the FIRE-channel program.

Each of the dialer's channels can be triggered by a switched dc voltage, a dry contact closure, or a dry contact opening. The trigger inputs may be either momentary or sustained. In either case, the dialer transmits its full program, then stops and resets itself. An input that is still present when the dialer stops must be removed briefly and then applied again to restart transmission on that channel. A sustained input does not make the dialer transmit or interfere with normal use of the telephones, nor does it interfere with triggering and operation of the dialer on its other channel.

When available, an appropriate dry contact closure should be used instead of a switched voltage for the dialer-trigger input. Figure 2-12 shows the preferred connections for a typical telephone dialer.

Where the contacts of a police-connect panel are needed for polarity reversal, the contacts may be used to provide a switched-voltage trigger for the dialer as shown in Fig. 2-13. This hookup lets the panel's BELL TEST feature be used without causing any dialer transmission.

When using the bell output of an alarm panel as a switched-voltage trigger for the dialer, always run the trigger wires directly from the dialer input terminals to

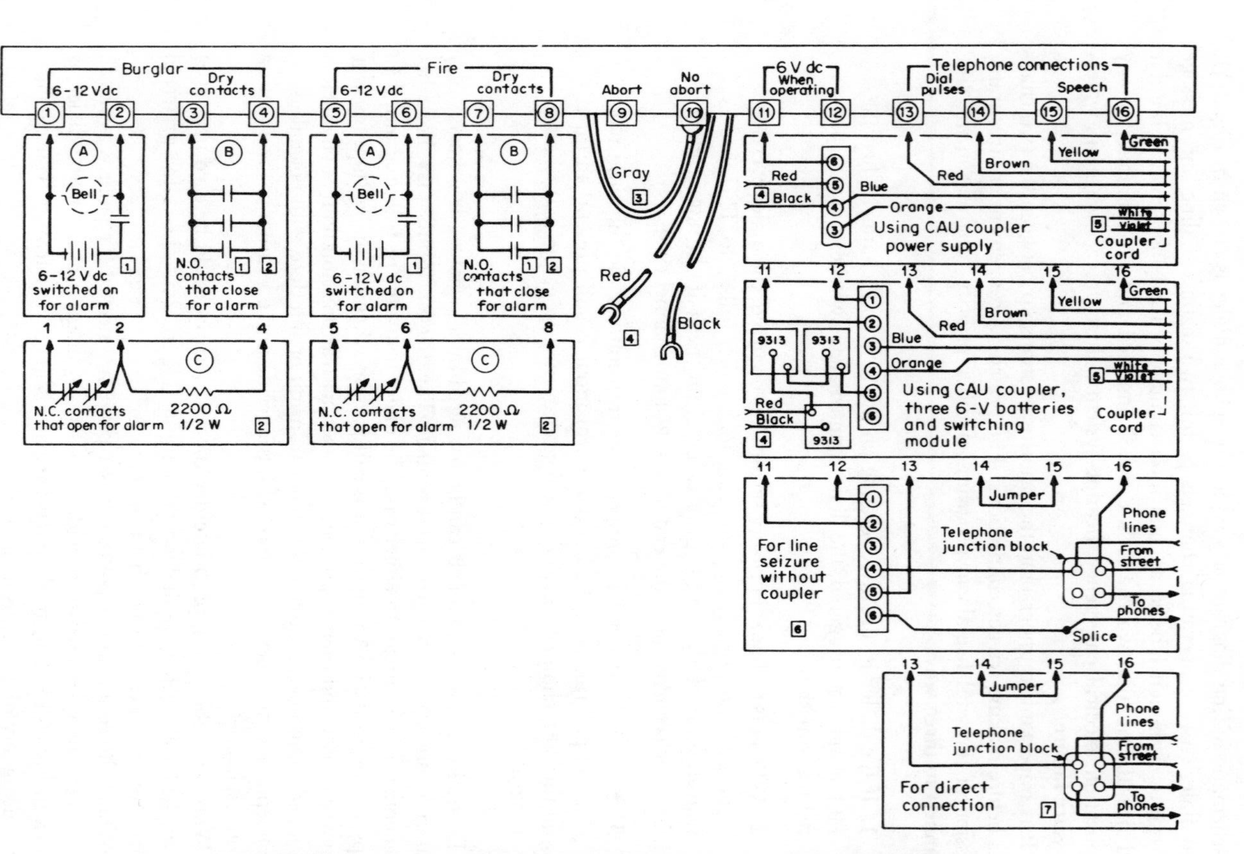

FIGURE 2-11. Schematic wiring diagram of a typical telephone dialer.

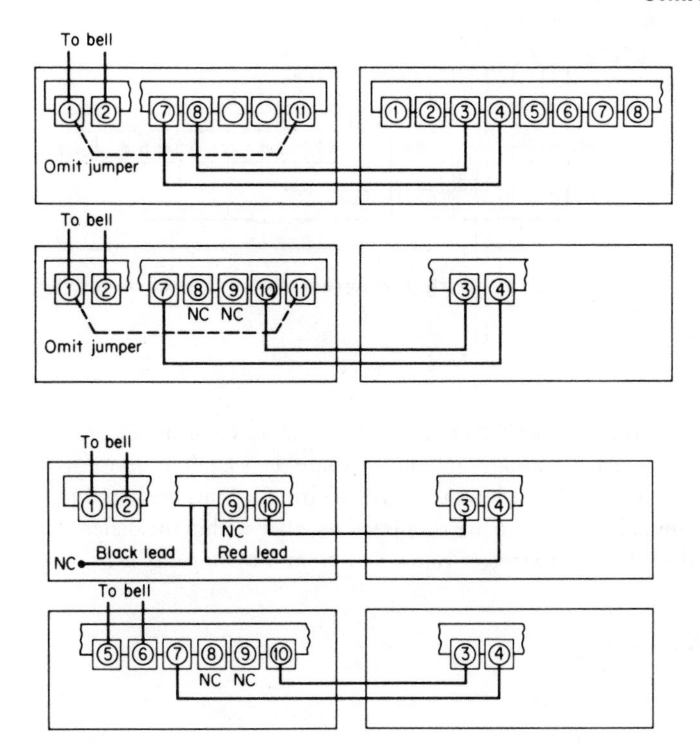

FIGURE 2-12. Preferred connections for a typical telephone dialer.

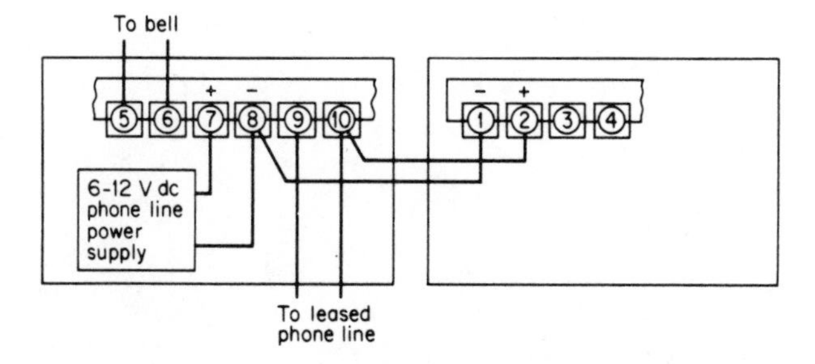

FIGURE 2-13. A switched-voltage trigger connected to a telephone dialer.

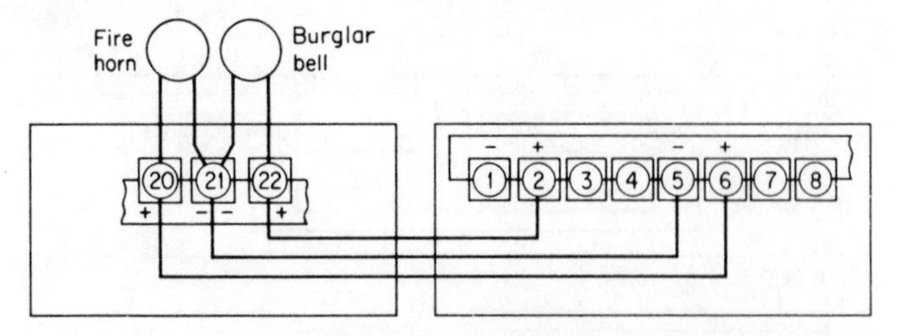

FIGURE 2-14. Wiring diagram for a switched-voltage trigger.

control the panel terminals. Do not run the wires from the dialer inputs to the bell, horn, or siren locations, and do not route the sounding-device wires through the cabinet. Figure 2-14 shows the correct wiring for this hookup. In this hookup, dialer terminals 2, 5, and 6 are connected together within the dialer. This permits a simplified three-wire trigger connection from the control panel.

3

BASIC INSTALLATION TECHNIQUES

Before the installation of a security/fire-alarm system is started, a sketch of the building should be prepared or the original blueprints should be obtained. This sketch should be drawn to scale and should show the location of all windows and doors, chases, closets, etc. A simple riser diagram showing the various components such as smoke and heat sensors, control panel, and alarm signals should also appear on the sketch. When this is completed, the installer can begin the design of the security/fire alarm system (see Chaps. 5 and 6).

INSTALLATION BASICS

The installation of a protective security/fire-alarm circuit should always start at the protective-circuit energy source, as if it were an end-of-line battery—a battery remote from the control panel—even though it may actually be a power supply installed in the panel. A pair of wires are run from this power source to the first contact location, but just the positive wire is cut and connected to the two contact terminals as shown in Fig. 3-1. The neutral or common wire is not cut, but continues on in parallel with the positive or "hot" wire. The pair is then run on to the next contact—be it door, window, sensor—and again only the hot wire is connected to the contacts. This procedure is repeated until all contacts are wired in series, and then the pair of wires is run from the last contact device on the system to the protective-circuit terminals in the panel. Although the markings will vary from manufacturer to manufacturer, the terminals for the starting connections will read something like LOOP POWER OUT, while the terminating terminals will read IN or a similar term.

A simple circuit of the wiring connections just described is shown in Fig. 3-2. Obviously, the system would operate with just a single-wire, positive-leg circuit run from contact to contact, with the negative power-supply terminal connected

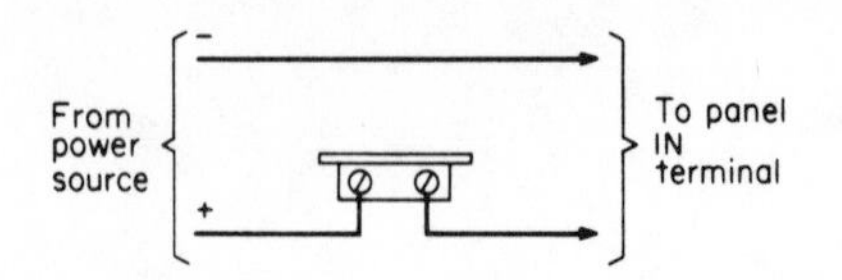

FIGURE 3-1. Contacts are connected into the positive wire only. Break positive wire only at door contacts.

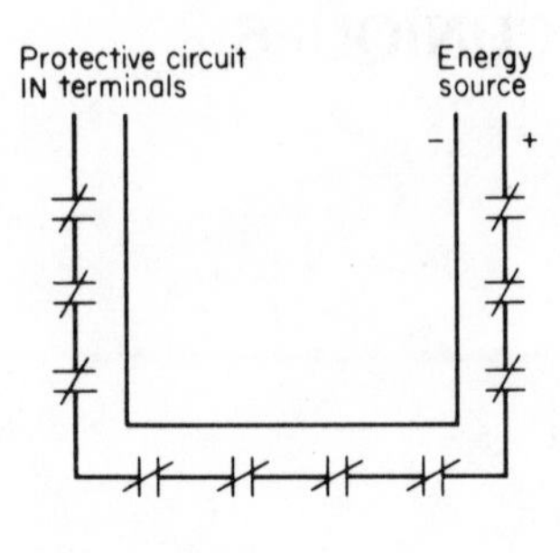

FIGURE 3-2. Negative conductor is run with positive conductor to all contacts even though the system would operate with just a single-wire, positive-leg wire run from contact to contact.

directly to the negative protective-circuit terminal within the cabinet. However, manufacturers discourage this practice, since troubleshooting a single-wire circuit can be extremely time-consuming and the single wire is more vulnerable to defeat by an intruder with no trouble symptoms occurring to warn the user of the loss of protection.

An exit/entry delay relay is sometimes used on security systems so that authorized personnel may exit and enter (using their door keys) without activating the alarm. However, a shunt switch is more often preferred (see Fig. 3-3). The purpose of the shunt lock is to enable an authorized person with a key to shunt out the contacts on the door used for entry/exit, allowing him or her to enter or leave the premises without causing an alarm when the alarm system is turned on. The

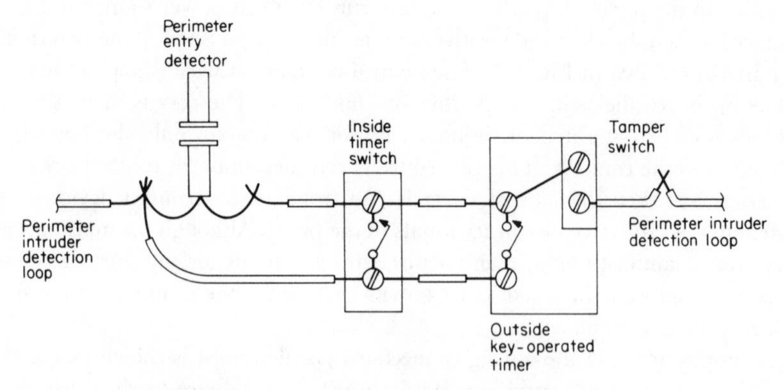

FIGURE 3-3. Typical shunt switch circuit.

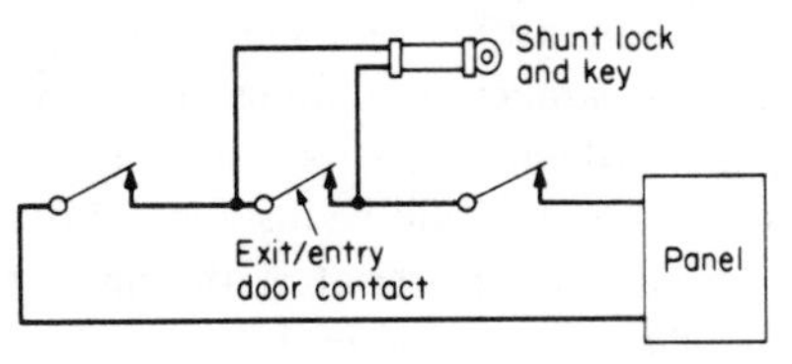

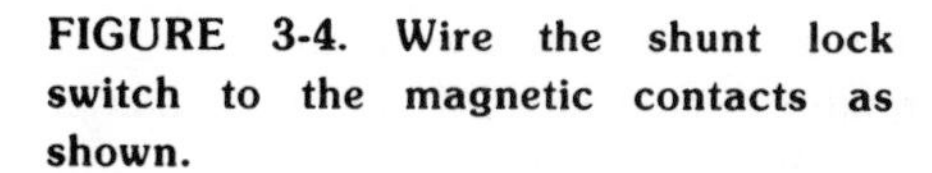

FIGURE 3-4. Wire the shunt lock switch to the magnetic contacts as shown.

shunt lock does extend outside the protected premises, however, and it is a potential weak link in the system. Following the two procedures suggested below makes defeat of the shunt lock much more difficult.

1. Install the shunt lock at the door that is most brightly illuminated and most readily visible to passersby.

2. Wire the shunt lock switch to the magnetic contact terminals as shown in Fig. 3-4. This arrangement traps the lock, so that any attempt to pull it out to gain access to its terminals will break the positive side of the protective circuit and cause an alarm to sound.

Contacts used to signal the opening of doors, windows, gates, drawers, etc., are usually mounted on the frame of the door or window, while the magnet unit is mounted on the door or window (moving part) itself. The two units should be positioned so that the magnet is close to and parallel with the switch when the door or window is closed. This keeps the shunt lock actuated, but opening the door or window moves the magnet away and releases the switch mechanism.

As long as the faces of the switch and magnet are parallel and in close proximity when the door or window is closed, they may be oriented side-to-side, top-to-top, or top-to-side. Mounting spacers may be used under the units if necessary to improve their alignment and proximity.

Terminal covers are available for most makes of door contacts to give the installation a more finished look and also to protect the terminal connections against tampering.

The wiring of any alarm system is installed like any other type of low-voltage signal system; that is, one must locate the outlets, furnish a power supply, and finally interconnect the components with the proper size and type wire.

QUALITY OF WORKMANSHIP

Since most security/fire-alarm systems are operated on low-voltage circuits, many installers might not pay as strict attention to the quality of the workmanship and materials as they would when installing conventional electrical wiring for lighting and power. Security/fire-alarm systems are worthy of the best materials and the best workmanship and strict attention to quality work should always be given.

Care must be taken to ensure that all visible components are installed adjacent to and parallel to building lines to give a neat appearance. All wiring should be concealed where possible, and the wiring that must be exposed should have square corners and should be installed so that it is as inconspicuous as possible.

Only new material of the highest quality should be used and this material should be approved by UL or a similar testing agency. Remember that the protection of the owner's building and its contents are dependent—to a great extent—on the quality of the security/fire-alarm system installed.

INSTALLING SYSTEMS IN EXISTING BUILDINGS

Many changes and advances in developing complete security/alarm systems for building operation and protection have taken place in the past few years. Numerous existing buildings are currently having security and fire-alarm systems installed—either to replace their obsolete systems or to provide protection they never had.

The materials used for installing a complete alarm system in an existing building are essentially the same as those used in new structures. However, the methods used to install the equipment and related wiring can vary tremendously and require a great deal of skill and ingenuity. Each structure is unique.

When concealed wiring is to be installed in a finished existing building, the installation must be planned so that the least amount of cutting and patching is necessary. In most cases, this means giving special consideration to the routing of conductors. Unlike the wiring of a new building where the installer would try to conserve as much material as possible, the amount of material used (within reason) is secondary in existing buildings. The main objective in security/fire-equipment installations in existing buildings is to install the wiring in the least amount of time with the least amount of cutting and patching of the existing finishes of the building.

Prior to any actual work on an existing building, the contractor or his installers should make a complete survey of the existing conditions in the areas where the security system will be installed. If the majority of the work can be done in exposed areas (as in an unfinished basement or attic), the job will be relatively simple. On the other hand, if most of the wiring must be concealed in finished areas, there are many problems to be solved. The initial survey of the building should determine the following:

1. The best location for the alarm control panel.

2. The type of construction used for exterior and interior walls, ceilings, floors, etc.

3. The location of any chases that may be used for routing the conductors and the location of closets, especially those located one above the other, for possible use in fishing wires.

4. The material used for wall and ceiling finishes—plaster, drywall, paneling, etc.

5. Location of moldings, baseboards, etc., that may be removed to hide conductors.

6. Location of decorations or other parts of the building structure that cannot be disturbed.

7. Location of any abandoned electrical raceways that new alarm-system wires might be fished into. Don't overlook similar possibilites. For example, old abandoned gas lines were recently used to fish security-system wires in an old building in Washington, D.C.

8. The location of all doors and windows, coal chutes, and similar access areas to the inside of the building.

As indicated previously, the most difficult task in running wires in existing buildings is the installation of concealed wiring in finished areas with no unfinished areas or access to them in the area in question. In cases like these, the work is usually performed in one of two ways, namely, by deliberately cutting the finished work so that the new wiring can be installed. Of course, these damaged areas must be patched once the wiring is installed. The second way is to remove a small portion of the finished area (only enough to give access to voids in walls, ceilings, etc.) and then fish the wires in. The removed portions are then replaced after the wiring is complete.

Where outlet boxes are used, they should be designed for installation in the type of finish in the area. Means of securing the boxes to some structural member—like mounting ears or holding devices—should also be given consideration.

Another method of providing outlets in a finished area is to remove the existing baseboard and run the conductors in the usual groove between the flooring and the wall and then replace the baseboard. This method requires less work (cutting and patching) than most other methods when the finished area must be disturbed. There is also a type of metal baseboard on the market which may be installed along the floor line and used as a raceway. Most types are provided with two compartments for wires—one for power and one for low-voltage wiring. Using this metal baseboard provides a simple means of routing wires for security/fire-alarm systems with very little cutting or patching. In most cases, wires can be fished from the baseboard up to outlets on the wall, especially if they are under 3 ft. (0.9 m) above the floor. However, if this is not practical, matching surface molding can be installed to blend in very nicely with the baseboard.

When a lot of cutting and patching is required in a finished area, many installers like to hire a carpenter to do the work. The carpenter may know some tricks that will help the alarm-system installers get the system in with the least amount of difficulty. Also, any cutting or patching will be done in a professional manner.

Before doing any actual cutting of an existing building to install security/fire-alarm components, the installer should carefully examine the building structure to ascertain that the wires may be routed to the contacts and other outlets in a relatively easy way. It is possible that a proposed outlet location, for example, could be moved only a foot or two to take advantage of an existing chase. Perhaps a smoke detector or similar component was originally located in a ceiling with insulation, which would make the fishing of cables very difficult. If the detector could be located on a ceiling containing no insulation, the job would be greatly simplified.

When cutting holes in ceilings for outlets, a drop cloth or paper should be spread underneath to catch all dust and dirt. Sometimes an old umbrella can be opened and hung upside down under the spot in the ceiling where the hole is being made to catch the debris and keep it off the rugs and furniture.

Holes for wires and components can be cut through plaster with a chisel, through wood with a keyhole saw after first drilling two or four pilot holes, and in brick or other masonry with a masonry chisel or rotary hammer. To locate the exact spot to cut these openings, it is best to first cut a very small hole in the center of the spot where the larger one will be made. This hole may then be used to locate the area between studs or—in the case of very old homes—the cracks between the plaster laths. It is then possible to shift the mark for the outlet openings so that all obstacles can be avoided and to provide proper anchoring of the outlet box or component.

There are a number of ways to pull and fish wires into walls and openings in finished buildings and, with a little ingenuity and careful thought, workers should be able to solve almost any problem of this kind that they may encounter.

When pulling wires into spaces between the joists in walls, a flashlight placed in the outlet box hole is often a great help when feeding the wires in or catching them as they are pushed near the opening. Under no circumstances should a candle or other open flame be used for this purpose. If one must see farther up or down the inside of a partition, a flashlight and mirror used in combination as shown in Fig. 3-5 is a great help. Many installers like to make their own mirror

FIGURE 3-5. A flashlight and mirror used in combination are useful for viewing conditions inside of partitions.

by gluing a small 2- x 3-in (5- by 8-cm) compact mirror on a handle resembling a wooden tongue depressor. Any type of small flashlight may be used.

Where it becomes necessary to remove floor boards during a security/fire-alarm installation, it should be done with the greatest of care so that the edges are not split. Split edges make a poor appearance on the finished job when the boards are replaced. Special saws may be purchased for cutting into floors or other surfaces without drilling holes to start the saw. Then if the tongue (on tongue-and-groove boards) is split off with a thin sharp chisel driven down in the crack between the boards, the board from which the tongue was removed can be pried up carefully without damaging the rest of the floor.

NEW INSTALLATION TECHNIQUES FOR EXISTING STRUCTURES

A few years ago, the Diversified Manufacturing and Marketing Co. (Burlington, NC 27215) patented a system which attaches a drill bit to a long flexible spring steel shaft and is known as D'versiBit. This system makes it possible to manipulate easily a drill bit in walls to accomplish complex installation maneuvers in existing buildings. The D'versiBit can be inserted into the wall cavity through a small opening and positioned accurately for drilling from midwall to attic or basement, from windows and doorways to basement or attic, etc. The development of this system makes penetration and cable retrieval a much simpler operation than it used to be. Following is a list of tools available for use with the D'versiBit system.

Bits: The three types of bits available for this system are shown in Fig. 3-6. The auger bit (Fig. 3-6*a*) is for starting and drilling a clean entrance hole, the combination bit (Fig. 3-6*b*) is designed for greater durability, and the masonry bit (Fig. 3-6*c*) has a carbide tip for drilling in cement blocks and plaster. All three of these bits are designed for use with standard drill motors.

Alignment Tool: The special alignment tool shown in Fig. 3-7 provides total control of the flexible shaft, and may be used to hold the bit and shaft steady and true toward any desired destination.

Line Recovery Devices: After the drilling is completed, the system quickly converts to a line recovery system using the grips as shown in Fig. 3-8. These grips attach to holes located in the bit tip or in the shaft end. This feature enables even one person to quickly fish wires or cables through partitions.

Shaft Extensions: The standard lengths of the flexible shaft are 54 in (135 cm) and 72 in (180 cm), but shaft extensions (Fig. 3-9) are available to provide extra distance drilling capabilities. One or more can be attached in special situations, such as from the basement to a smoke sensor in the attic.

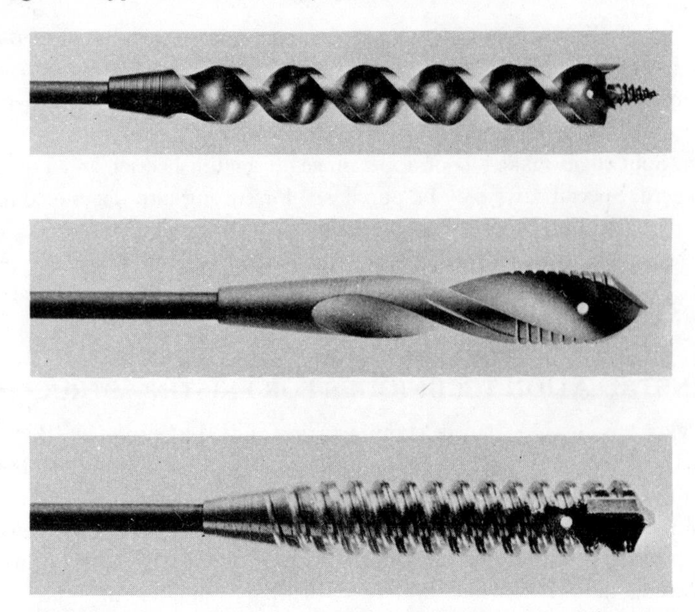

FIGURE 3-6. Three types of bits available for the D'versiBit system.

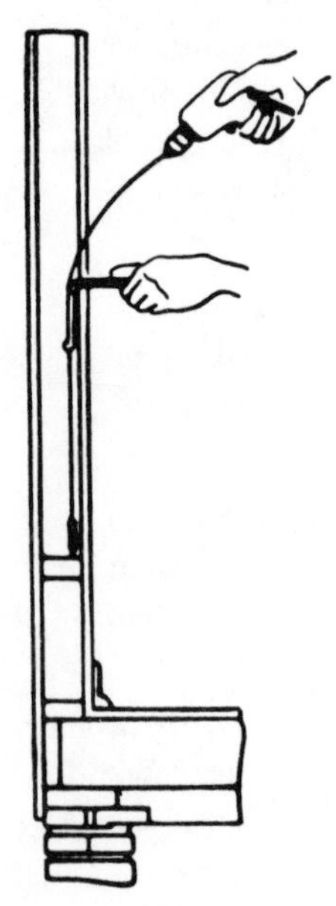

FIGURE 3-7. Special alignment tool provides total control of the flexible shaft.

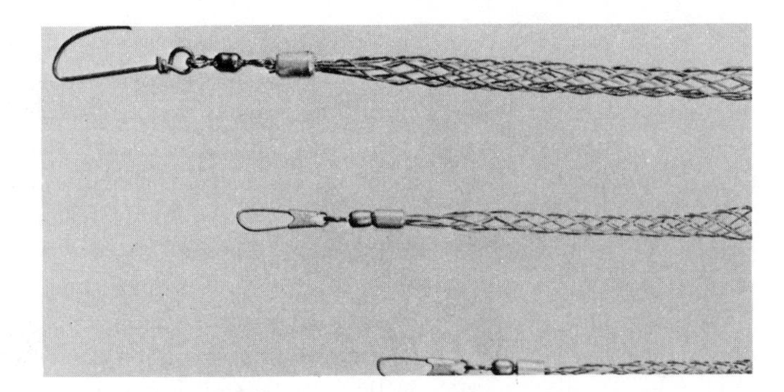

FIGURE 3-8. Line recovery devices.

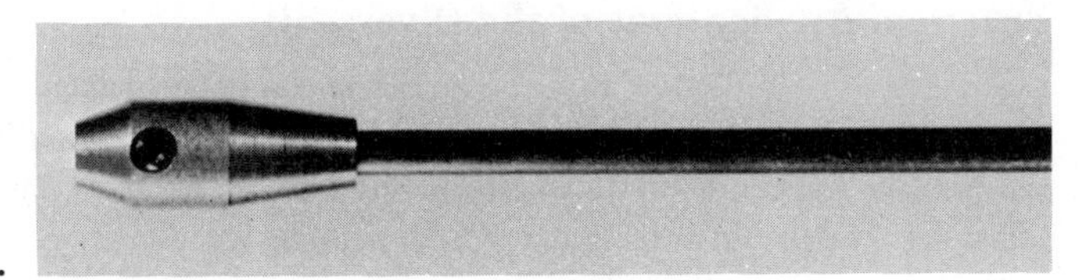

FIGURE 3-9. Shaft extension.

The basic shaft is 3/16 in (0.5 cm) which will accommodate both 3/8-in (0.9 cm) and 1/2-in (1.3-cm) drill bits in the three styles mentioned previously. For larger bits—such as 3/4 in (1.9 cm) and 1 in (2.5 cm) sizes—a 1/4-in (0.6-cm) shaft is required. This larger shaft reduces the flexibility for complex drilling.

Operation Procedures

When drilling with the flexible shaft of the D'versiBit, run the drill motor only when actually drilling. Never run the drill when sliding the bit up or down in the wall cavity as wires—either signal wires or existing electric power wiring—may be cut during the process. Also make certain that the bit is sharp since a dull bit is one of the greatest causes of bit breakage.

If at all possible, a reversible drill motor should be used to withdraw the bit from the wall. The motor should be running only when the bit is actually passing through a wood member. When drilling, force is exerted in one direction. When the bit is being removed, it is removed at a different angle and force is exerted from a different direction. This is why the reverse is used. If the flexible shaft is being used with drill motors with no reverse, it would be better to exert force to pull the bit from the hole with the motor running, because chances of an easy recovery without damage are much better with the motor running.

When drilling from an attic or crawl space, be certain not to select an area directly above or below a door since this will result in property damage. It is also good to keep a slight tension on the wire when it is being pulled from overhead so that it will not get tangled with the bit and become damaged.

The shaft should not be bowed any more than absolutely necessary to accomplish the job. Excessive bowing will decrease the life of the flexible shaft. Drill motors, of course, should be adequately grounded or else have insulated handles.

Practical Applications of the D'versiBit

Assume that an outlet box for an infrared photoelectric detector is to be installed above a countertop in a residential kitchen to sense entry of unauthorized persons through the kitchen door. If, upon investigation of the space inside of the partitions, it is found that a 2- by 4-in (5- by 10-cm) wood member (fire-stop) blocks the route from the outlet hole to the basement area where the alarm control station is located, an alignment tool must be used.

The flexible shaft containing a drill bit is placed through a cut outlet-box opening and then the special alignment tool is attached to the shaft as shown in Fig. 3-10. By keeping the alignment tool in the same position on the shaft and by lifting the handle, the shaft will bow back toward the operator. As the bit is lowered into the wall cavity, the operator can feel the bit strike the inside wall. When the bit is aligned correctly on the wooden member, the alignment tool is removed while keeping downward pressure on the bit so that it will not slip out of place, and the hole is drilled through the fire-stop. This hole will then act as a guide for drilling through the floor plate as shown in Fig. 3-11.

In the case of a wall cavity without fire-stops or purlins, the alignment tool is used to snap the bit back to the inside wall (Fig. 3-12) at which time downward pressure on the drill motor will keep the bit point in place and cause the shaft to bow. Power and pressure is then transmitted from the back wall which allows proper angle drilling to miss the joist boxing.

After the bit has penetrated into the basement area as shown in Fig. 3-13, the operator has access to the hole in the drill bit itself for attaching the recovery grip and pulling the wire up to the outlet location—all without damage to existing finishes.

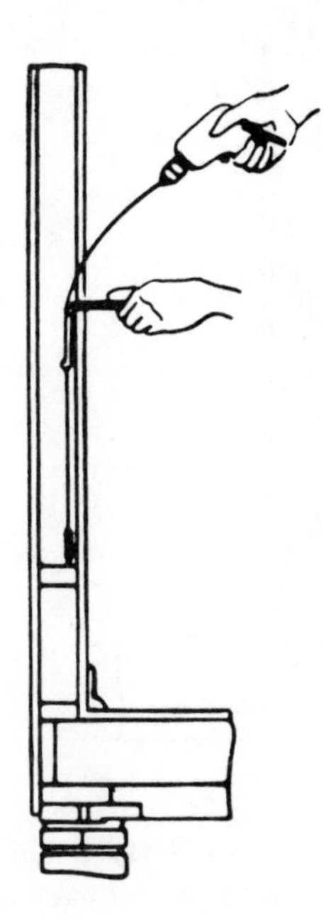

FIGURE 3-10. The alignment tool is attached to the shaft ready for operation.

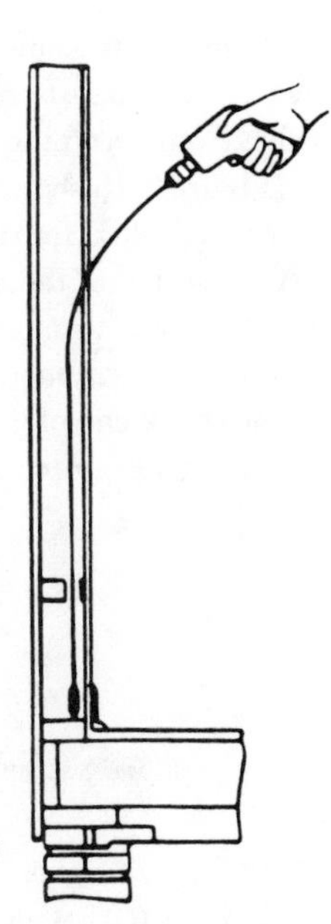

FIGURE 3-11. The first hole cut acts as a guide for drilling through the floor plate.

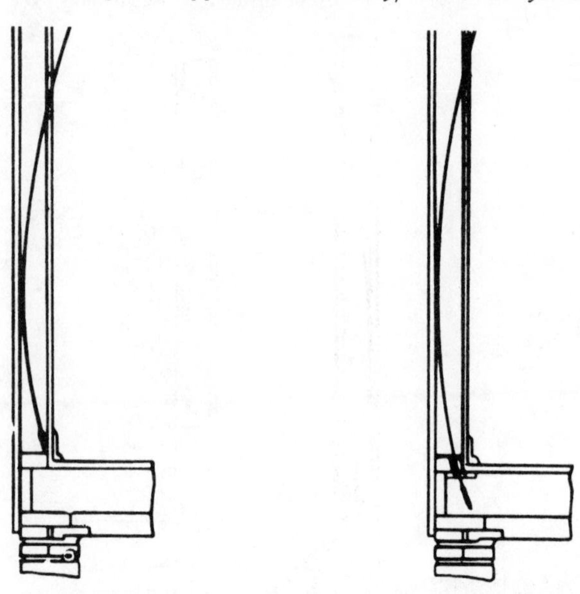

FIGURE 3-12. Alignment tool used to snap the bit back to the inside wall.

FIGURE 3-13. Bit has penetrated into basement area.

Figure 3-14 shows how the recovery grip is attached to the bit tip eyelet. The swivel located between the cable and the head of the grip prevents the wire or cable from becoming twisted during the fishing process.

Figure 3-15 shows the grip after it has been attached to the bit tip with the line inserted ready for recovery. The operator then operates the drill motor in reverse—due to the angle of the pull—applies a slight pull, and the wire can be pulled easily through the holes due to the reverse cutting action of the bit. If desired, the drill motor can be removed from the shaft and a recovery grip attached to the chuck end of the shaft for pulling the wires downward toward the basement. While this example shows the method of routing wires or cables from an outlet

FIGURE 3-14. Recovery grip attached to the bit tip eyelet.

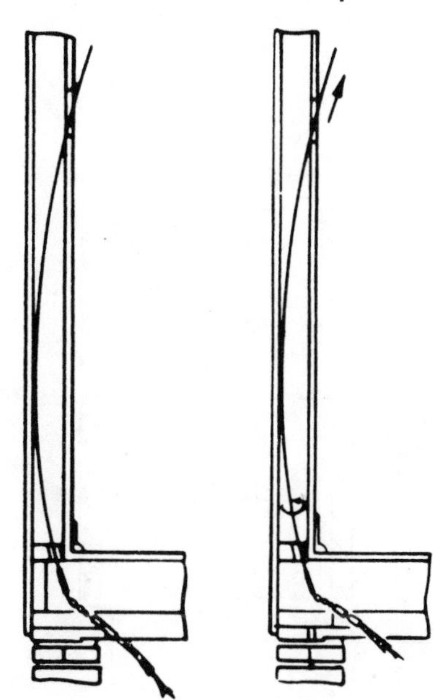

FIGURE 3-15. Grip attached to the bit tip with the line inserted ready for recovery.

to a basement, the same procedure would apply for drilling from an outlet opening to an attic space.

To install contacts on windows for a burglar-alarm system, drill from the location of the contact through the casement, lintels, and plates with a ⅜-in (0.9-cm) shaft. Attach a recovery grip to the end of the bit, insert the wire to keep the grip from becoming tangled, reverse the drill motor, and bring the wire toward the operator as the bit is being withdrawn.

Burglar-alarm contacts or door switches installed at doors are simple projects when one uses the flexible shaft. First cut or drill the entrance hole in the normal manner and then insert the flexible shaft with bit into the entrance hole, slanting the bit as much as possible in the desired direction of travel. Continue by drilling through the door casing and floor jamb into the cavity of the wall as shown in Fig. 3-16. The drill is then stopped until it strikes the next stud which will deflect the bit either up or down, depending on the direction of the drilling. Continue to push the bit until it strikes the top of the bottom plate and then drill through the plate into the basement or attic. The recovery grip is then attached to the bit and the wire or cable may be drawn back toward the operator by reversing the drill motor and keeping a slight tension on the wires as they are being pulled to prevent tangling.

With conventional tools, the routing of wires from one outlet to another—as shown in Fig. 3-17—requires either channeling the wall, using wire mold, or

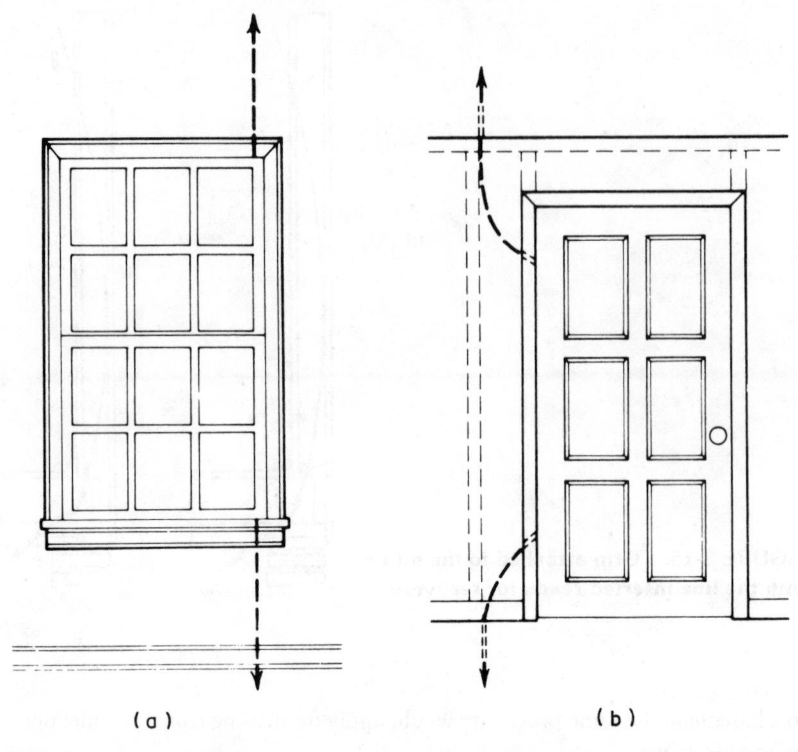

FIGURE 3-16. Drilling through the window casing (*a*) and door jamb (*b*) into the cavity of the wall.

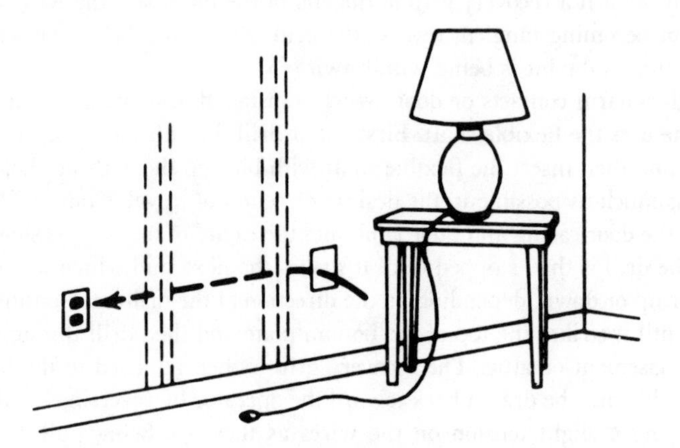

FIGURE 3-17. How wires must be routed when one uses conventional tools.

running the wires down to the baseboard, removing the baseboard, and then installing the wires behind it. Instances like these occur when the crawl space is too shallow for workers to crawl into or the house is built on a concrete slab. However, with the flexible shaft, it is possible to drill through the wall horizontally through several studs (if the operator is careful) and then pull the wires back through the holes to the openings.

The installation of an outside annunciator under the eave of a house with an extremely low pitch to the roof would cause several problems in getting wires to the outlet. With the flexible shaft, a hole can be drilled through the boxing as shown in Fig. 3-18. As soon as the bit penetrates the boxing, it is pushed into the attic as far as it will go. A recovery grip is then attached to the bit, the wire or cable inserted, and then pulled backward toward the outlet opening. The outlet box and annunciator (horn, bell, etc.) are installed under the eave and the other end of the cable is connected to the alarm system. Also, because the flexible shaft is more rigid than the conventional fish tape, it will penetrate attic insulation if any exists.

If it becomes necessary to install wiring in an attic and run cable from this area to the basement, the installation can be greatly simplified by using a flexible shaft. First drill through the top plate into the wall cavity—making sure that the drilling is not being done above a window or doorway or any other obstruction such as existing wiring, ductwork, etc. Once through the top plate, the drill motor is turned off and the bit is pushed into the cavity of the wall as far as it will go. If no fire-stops are encountered, the bit is pulled back and an extension is attached to the shaft. With the extension installed, the bit is again lowered into the wall cavity until a fire-stop is encountered. The bit is then positioned and used to drill through the wooden member. Once the wooden member is penetrated, the drill

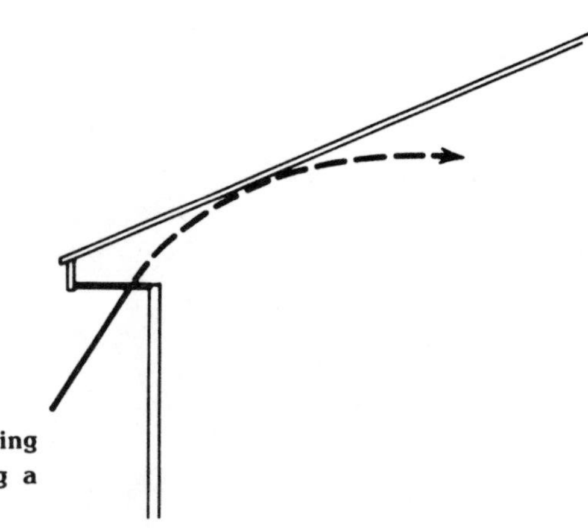

FIGURE 3-18. Method of drilling a hole through boxing by using a flexible shaft.

motor is again stopped and the bit is lowered further until the bottom plate is reached. Continue drilling through the bottom plate in the basement or crawl space. Fasten the appropriate recovery grip, insert the wire or cable, and pull up the wire with the flexible shaft. The drill motor should be reversed only when the bit is passing through one of the wooden members.

Those who use this device often are certain to discover many other useful techniques for installing wiring in existing structures.

4

SELECTING EQUIPMENT

Dozens of manufacturers in this country offer security/fire-alarm systems with a wide selection of accessories to fill practically any need. The selection of a particular system for a given application will usually depend upon the following factors:

1. The type of building to be secured.
2. The allotted budget.
3. The availability of the equipment.
4. Service available.

RESIDENTIAL EQUIPMENT

The diagram in Fig. 4-1 illustrates various components of a residential security/fire-alarm system as distributed by NuTone Housing Products. The following is a brief description of each component and its function within the system.

The surface magnetic detector is the most versatile entry detector for residential alarm systems and should be considered first as a method of protecting any movable door or window. These detectors can be mounted on wood, metal, and even glass, if necessary. They can be mounted with screws, double-sided tape, or epoxy. Obviously, the tape and epoxy are useful on glass, aluminum, or any other surface where screws cannot be used. However, when using tape or epoxy, make certain that the surface is clean, dry, smooth, and at least 65°F (18°C) when applied.

Where the appearance of surface-mounted systems is objectionable, recess-mounted magnetic protectors may be used. These detectors are more difficult to install—requiring greater care on the installer's part—but few problems develop if the following precautions are adhered to:

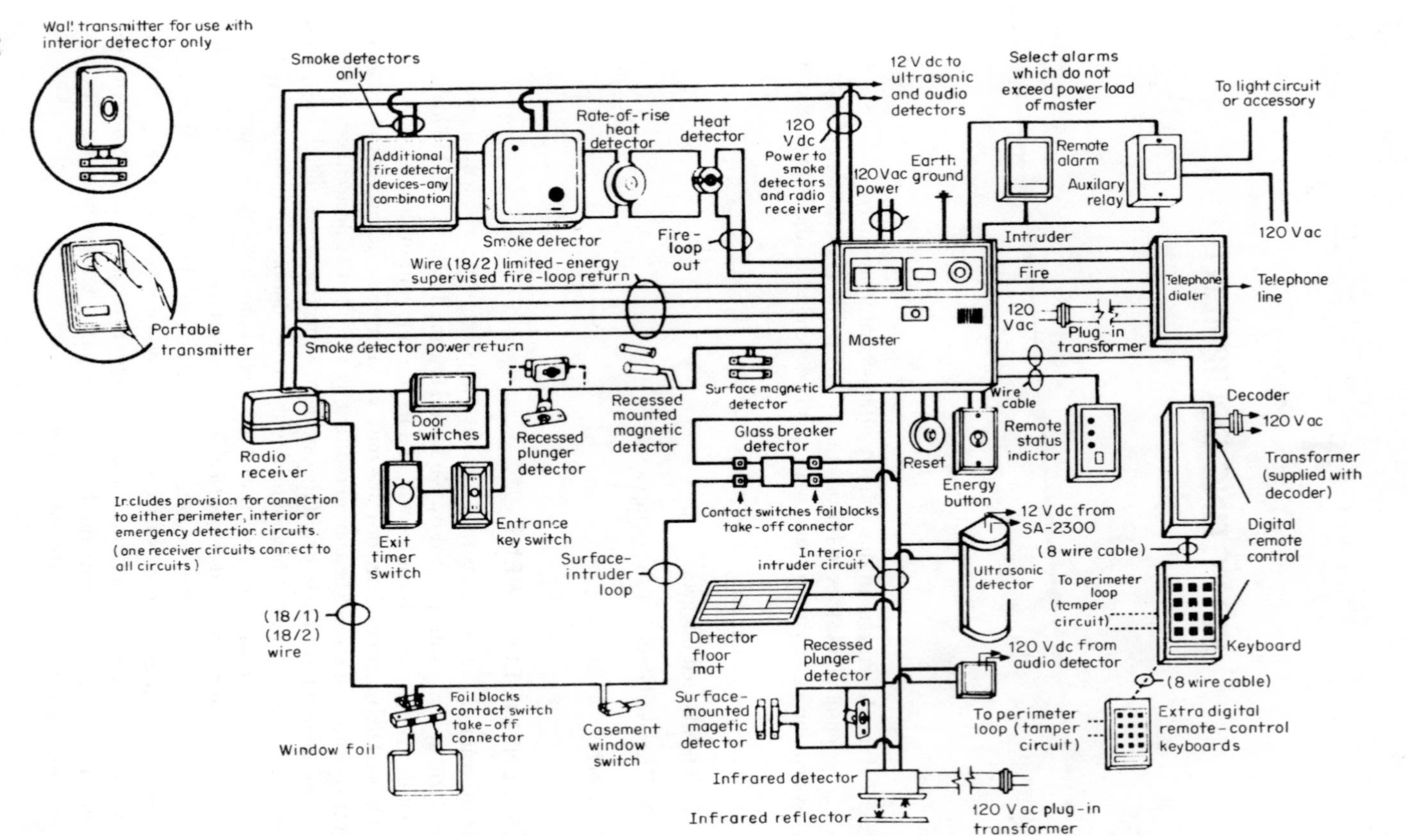

FIGURE 4-1. Various components of a residential security/fire-alarm system.

1. Be careful not to damage or destroy any weatherproofing seal around windows, doors, or other openings.

2. If a recessed-mounted entry detector is installed in the window sill, you must prevent water seepage to the switch by applying a sealant under the switch flange and around the switch body.

3. When drilling holes to accept each half of the detector, be sure the holes line up and there is no more than ¼-in (0.6-cm) space between the two sections of the detector.

4. Be certain there is enough space between the window and its frame (or door and its frame) when each is closed; that is, there must be enough space (usually equaling 1⁄16 in or 0.16 cm) for the protrusion of both sections when they meet.

5. If the window frame is not thick enough to accept the magnetic section of the detector, the detector can be mounted in the side frame.

The recessed plunger detector shown in Fig. 4-1 is mounted so that the door or window will contact the plunger at the tip and push the plunger straight in. Therefore, the area of the window or door that depresses the plunger should have no slots, cutouts, or step-downs into which the plunger might slip. The area should also be hard and free of rubber or vinyl that might be weakened by the plunger and consequently allow the plunger to open. For protecting doors, plunger-type detectors should only be mounted in the door frame on the hinge-side of the door.

In cases where it is difficult to protect a window or door by mounting any of the direct-type detectors, the area directly inside the door or window can be protected with interior "space" detectors, such as a floor-mat detector (Fig. 4-2) or an ultrasonic motion detector (Fig. 4-3).

Floor-mat detectors are easily concealed under rugs at doors, windows, top or bottom of stairways, or any other area onto which an intruder is likely to step. A light pressure on the mat triggers the alarm.

There are also rolls of super-thin floor matting that can be cut to any desired length. These rolls can be used on stair treads and in areas near sliding glass doors or other large glass areas, entrance foyers, etc. In households with unrestricted

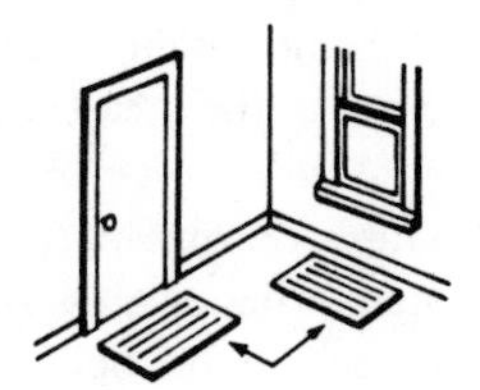

FIGURE 4-2. Floor-mat detector.

FIGURE 4-3. Ultrasonic motion detector.

pets, these mats are almost useless since the pets roam around the home and are certain to step on one of the mats and trigger the alarm.

Other space detectors include ultrasonic motion detectors, audio detectors, and infrared detectors. Care must be used with any of these units because the protected area is limited both in width and depth—depending upon the particular unit.

The ultrasonic motion detector can be used in large glass-walled rooms that might otherwise be difficult to protect and in hallways or entries or in virtually any area an intruder would have to pass through in moving about a home. They are especially useful as added protection (when conventional detectors are used also) to monitor a "valuables" room or area.

Most ultrasonic motion detectors are designed for mounting on either the wall or ceiling. It emits inaudible high-frequency sound waves in an elliptical pattern that ranges from 12 ft (4 m) to 35 ft (11 m) by 5 ft (2 m) by 20 ft (6 m) for most residential models. When an intruder moves within the secured area, movement interrupts the established pattern of sound waves and sounds the alarm.

Some designs of motion detectors can be rotated up to 180° for maximum coverage of the area being monitored as shown in Fig. 4-4.

Another type of motion detector is the *audio detector* (Fig. 4-5). This type senses certain sharp sounds known to be present in forced entry, such as wood splintering or glass breaking. When these sounds are received through the unit's miniature microphone, the detector triggers the control unit to sound an alarm.

Audio detectors are best utilized in areas which are seldom used, such as an

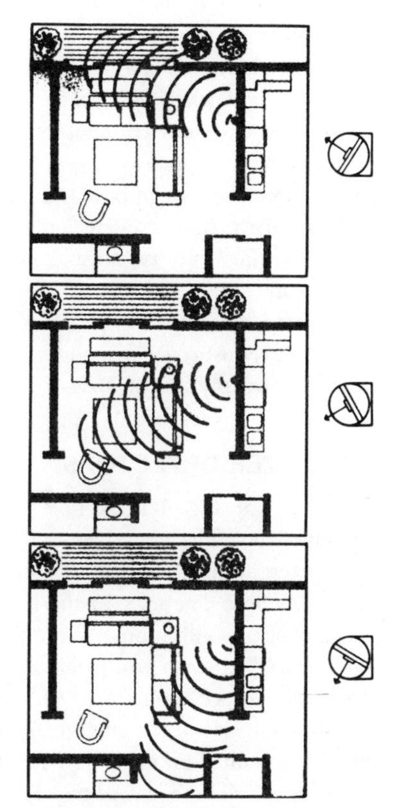

FIGURE 4-4. Motion detector rotating up to 180 degrees for maximum coverage of the area being monitored.

FIGURE 4-5. Audio detector.

attic, garage, or closed-off wing. It can be used in other areas, but when such areas are subject to much daytime activity, it is recommended that the detector only be armed at night when the family retires or is away from home.

Infrared detectors are another type of motion detector. A combination transmitter-receiver is used to project an invisible pulsating beam at a special bounce-back reflector on an opposite wall. Any interruption of the beam activates the system alarms. Infrared detectors can be wired to either the perimeter or interior circuit, but for faster response, it is recommended that it be connected to the interior circuit.

Infrared detectors are designed for indoor areas such as entries, hallways, rooms, etc. Most cover a span from 3 ft (1 m) to 75 ft (23 m), so it may be used in practically any indoor area or room.

PERIMETER DETECTORS

Refer again to Fig. 4-1 and note the various detectors available on the perimeter intruder loop. The glass-break detector, for example, is an excellent means of monitoring large areas of glass such as sliding glass doors, picture windows, and the like. These detectors, as the name implies, respond only to glass breaks and not to shock or vibrations. Therefore, they are relatively free from false alarms. The area which each will protect varies from manufacturer to manufacturer, but most will average about 10 ft^2 (0.9 m^2) of protection. A small cube like the one in Fig. 4-6 connects to the emergency circuit and the supervised perimeter circuit if they are mounted on movable windows.

Window foil tape is used mostly in commercial and industrial buildings but are sometimes used in residential systems—especially on basement windows. If an intruder breaks the glass, the tape tears, opening the circuit, and causes the alarm to sound.

Where the building construction makes it difficult to install wires, radio con-

FIGURE 4-6. Glass-break detector.

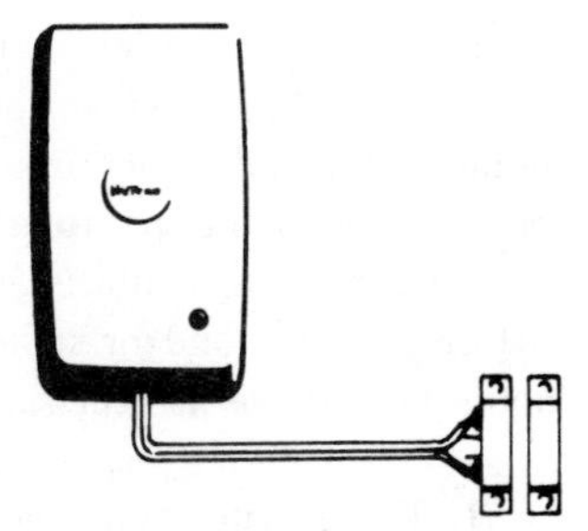

FIGURE 4-7. Wall-mounted radio transmitter.

trolled intruder detection systems are available. Such systems are also useful for linking outbuildings in a range of 150 ft (46 m) or more, depending on the type used.

Wall-mounted radio transmitters (Fig. 4-7) are easily mounted behind drapes at windows, above doors, and similar locations. Any number of transmitters can be used and each can be wired to an unlimited number of detectors as previously described.

When a detector senses forced entry, the transmitter sends a signal via radio waves to the radio receiver. It signals the control unit to sound an alarm.

FIRE-ALARM SYSTEMS

Most residential smoke detectors are photoelectric so that when abnormal smoke accumulates, they automatically activate the system alarms. Heat detectors are also used with smoke detectors in critical areas to help assure full coverage of the home. They activate the alarm when the preset temperature limit is exceeded, causing the contacts to close.

Another type of heat detector is sometimes referred to as a "rate-of-rise" type since it detects abnormal heat from either flash or slow-burning fires. Pneumatic rate-of-rise elements sense any rapid change in temperature, such as 12 to 15 degrees per minute, and sound the alarms. Also, if the fixed temperatures of the detectors are exceeded, the fusible alloy melts, closes the contacts, and activates the alarms.

COMMERCIAL SECURITY/FIRE-ALARM SYSTEMS

Security/fire-alarm systems used for commercial applications are similar to the ones previously described for residential use except that for the former, heavier-duty components are normally used along with additional equipment, such as automatic telephone dialers.

Magnetic contacts are used on doors and windows in closed-protective circuits, in direct-wire systems, and also in open-circuit applications. Movable elements within the switch unit of the magnetic contacts usually consist of a single flexible contact arm that provides a solid metal circuit path from the terminal screw to the contact-point end. The circuit continuity should not depend upon conduction across a hinge joint or through a coil spring.

When magnetic contacts are mounted on either noncoplanar or ferromagnetic surfaces, magnet and/or switch units should be held away from their respective mounting surfaces as necessary to:

1. Bring switch and magnet into close proximity when the door, window, etc., is closed.

2. Reduce the shunting effect of ferromagnetic materials so that positive switch pull-in occurs when the magnet approaches to within ⅛ in (0.3 cm) of the switch.

Mechanical contacts are used as emergency, panic, or fire-test switches. Ball contacts (Fig. 4-8) and plunger contacts (Fig. 4-9) are used in both closed- and open-circuit applications.

FIGURE 4-8. Ball contacts.

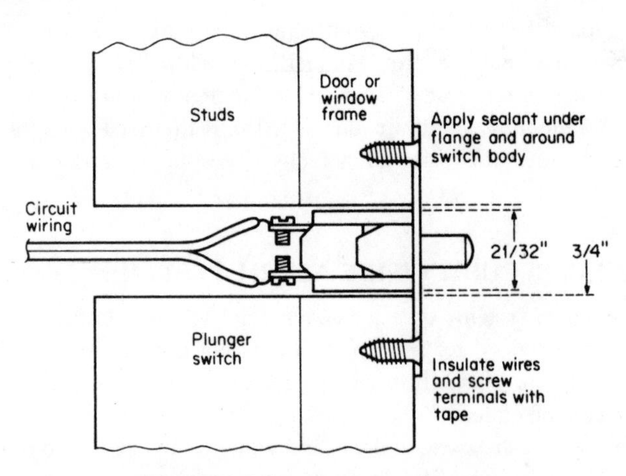

FIGURE 4-9. Plunger contacts.

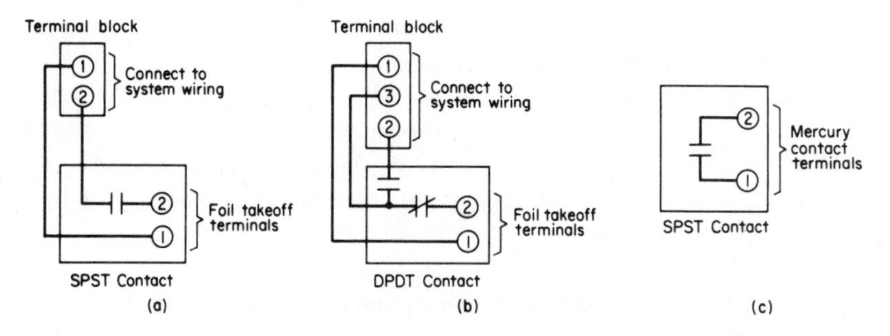

FIGURE 4-10. Wiring diagram of mercury contact connections.

Mercury contacts are sometimes used in low-energy alarm or signal systems to detect tilting of any horizontally hinged window, door, cover, access panel, etc. Due to the different items to be protected, it is best to install mercury contacts that can be adjusted to sensitivity after installation.

For combined detection of either opening or breakthrough, cord-mounted contacts with foil connected to takeoff terminals should be used. Wiring diagrams of mecury contact connections are shown in Fig. 4-10.

Holdup switches are usually installed under counters or desks in banks or stores, where an employee observing a holdup may be able to signal for help.

In banks and similar places where large amounts of money are exchanged, a money-clip alarm device is sometimes used. This device automatically triggers an alarm when all bills are removed from a cash drawer. A bill inserted in the clip (see Fig. 4-11) holds its switch in the normal position. Additional bills on top of the clip keep it concealed. Bills may be added or removed as required for normal

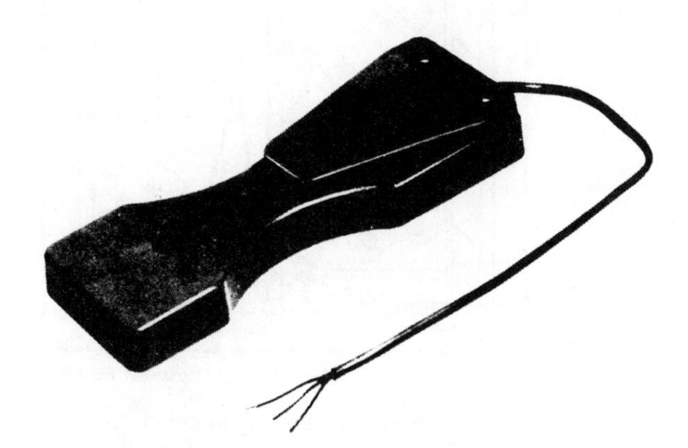

FIGURE 4-11. Money-clip alarm device.

business operations as long as one remains in the clip. However, the removal of all bills trips the clip switch to signal an alarm.

Money-clip alarm devices should be installed in the largest bill compartment of cash drawers and connected to the building alarm system by means of a retractable door cord. If exceptionally busy working conditions create the possibility of a false alarm since the bill in the clip might be accidentally removed, two money clips should be used at each station and wired so that both must be emptied to cause an alarm.

Window foil is used extensively in commercial applications. For fixed windows, the connections to the building alarm system is usually made through foil blocks as shown in Fig. 4-12. For movable windows and doors, a retractable door cord (Fig. 4-13) must be used.

Ultrasonic motion detectors for commercial applications are essentially the same as the ones described for residential use. However, the range of detection is sometimes extended on the units designed for commercial use. For example, a typical coverage pattern of a motion detector manufactured by Conrac is shown in Fig. 4-14. Note the coverage here is 15 ft (5 m) wide by 30 ft (9 m) deep.

Commercial telephone dialers are available that dial emergency numbers and deliver voice messages. Most distinguish between burglar and fire-alarm channels. A typical wiring diagram is shown in Fig. 4-15.

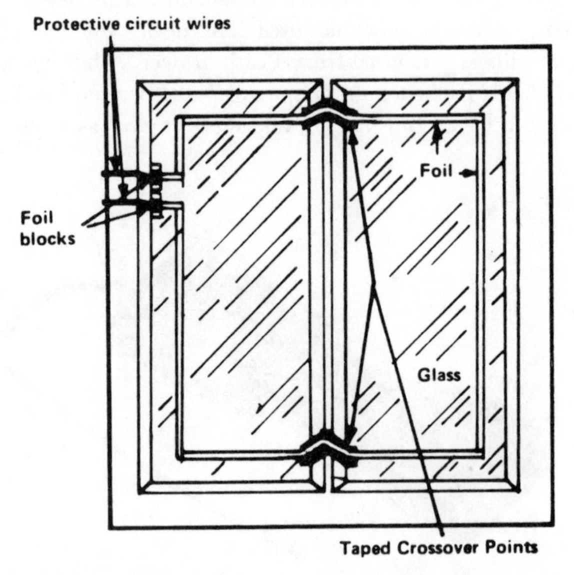

FIGURE 4-12. Window foil connected to foil blocks.

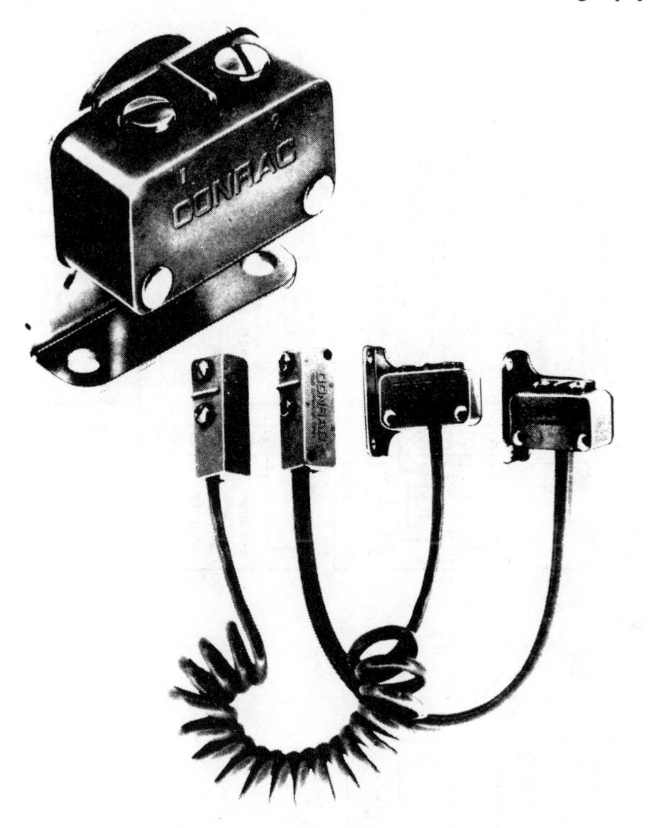

FIGURE 4-13. Retractable door cord.

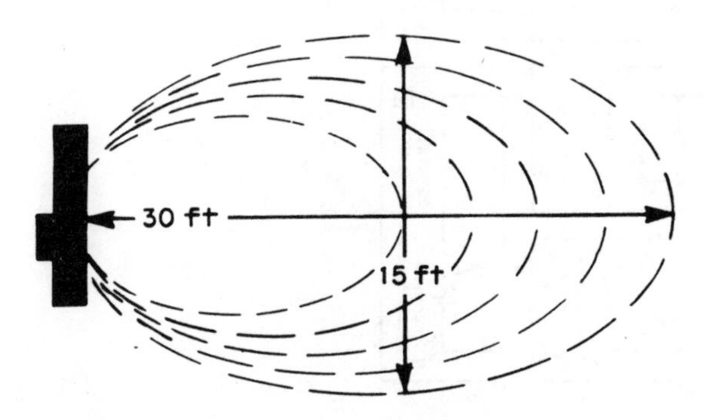

FIGURE 4-14. Typical coverage pattern of a motion detector.

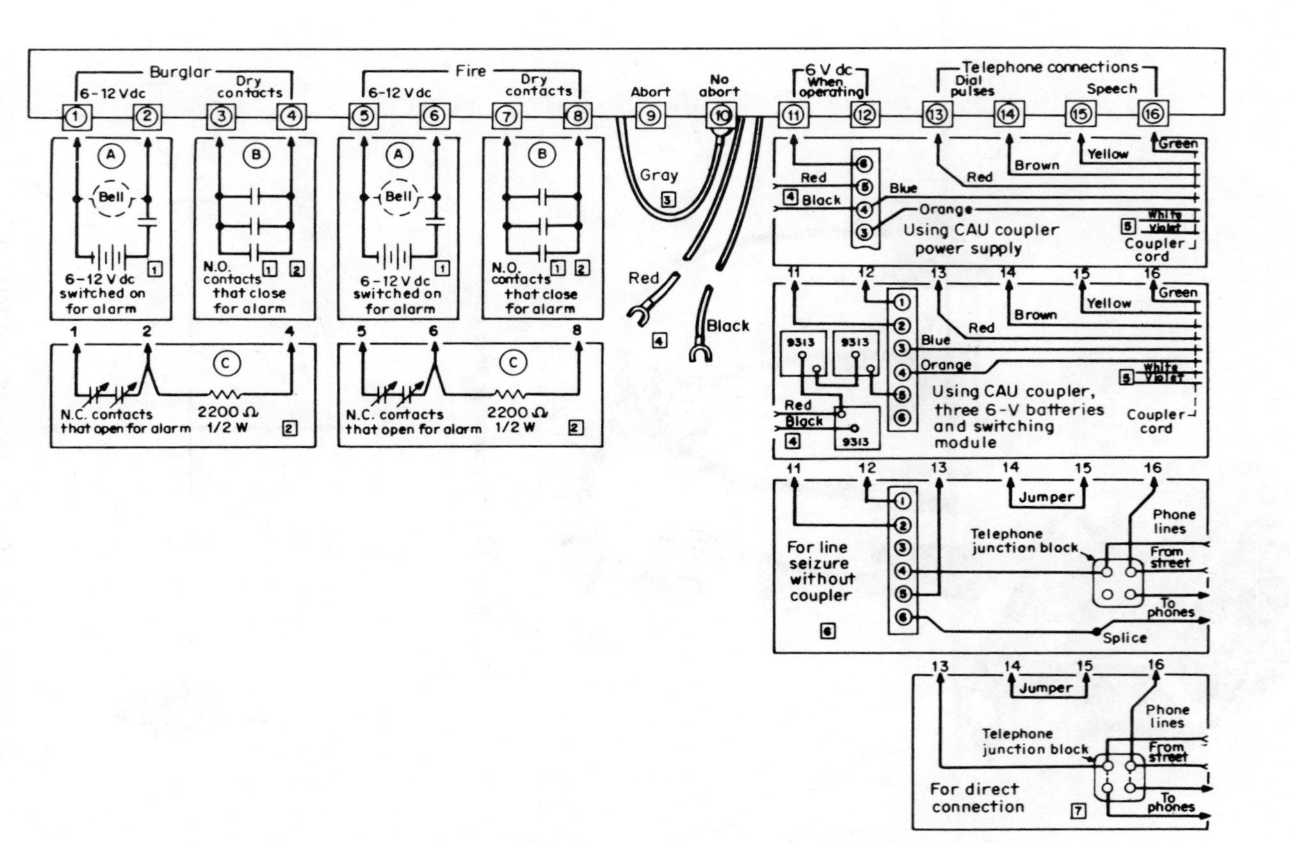

FIGURE 4-15. Wiring diagram of a commercial telephone dialer.

Digital alarm transmitters are becoming more popular for both commercial and industrial applications. They can be programmed on memory chips to meet the exact requirements of any business. Sample printing formats for one digital-alarm system manufactured by Adcor Electronics is shown in Fig. 4-16. In this model, each line (corresponding to an alarm code) is limited to 13 characters. A blank space between two words on the same line will take up one character.

These units are specifically designed for central-station monitoring of commercial and small industrial buildings. The unit consists of a transmitter, a special module, and a subscriber control station.

DEC 2 2 02 A M LOCATION 1111 FIRE HOLD-UP BURGLARY 1 BURGLARY 2 AUXILIARY OPEN CLOSED LOW BATTERY RESTORED	JAN 3 2 02 P M LOCATION 1111 FIRE HOLD-UP BURGLARY AUXILIARY 1 AUXILIARY 2 AUXILIARY 3 AUXILIARY 4 LOW BATTERY RESTORED
APR 4 5 05 P M LOCATION 1111 FIRE HOLD-UP BURGLARY 1 BURGLARY 2 SUPERVISORY OPEN CLOSED LOW BATTERY RESTORED	DEC 3 4 04 P M LOCATION 1111 POLICE FIRE SECURITY 1 SECURITY 2 CONTROL ARMED ROBBERY BURGLARY LOW BATTERY RESTORED
DEC 2 2 05 P M LOCATION 1111 BURGLARY FIRE PANIC FAILURE SYSTEM OFF SYSTEM ON AUXILIARY LOW BATTERY RESTORED	APR 7 6 26 P M LOCATION 1111 FIRE HOLD-UP NITE 1 NITE 2 SUPER DAY NORMAL NITE NORMAL LOW BATTERY RESTORED

FIGURE 4-16. Sample printing formats for digital alarm system.

The Adcor module has three input zones as follows:

Zone A: Two-wire, closed-circuit input generally used for a foil circuit. This zone may be 24-hour or key-controlled by the subscriber control station.

Zone B: Three-wire, closed-circuit and open-circuit input for perimeter doors and windows. This zone is armed and disarmed by the subscriber control station. It can be connected to the detector contacts so that an open or a crossed circuit will produce an alarm.

Zone C: Three-wire, closed-circuit and open-circuit input for internal protection devices, such as ultrasonic motion detectors, infrared beams, passive infrared detectors, etc. They may be wired so that either an open or a crossed circuit will produce an alarm. Six output leads are provided from the module to trip the control as follows:

1. Zone A open
2. Zone B open or crossed
3. Zone C open or crossed
4. System armed (closed—night)
5. System disarmed (open—day)
6. Alarm circuit restored

Zones A and B may be connected to the same channel of the transmitter, thus using only five channels for the module. If this is done, a daytime foil break will be reported as "perimeter alarm—day." The sixth channel may then be used for 24-hour reporting of fire alarm, holdup alarm, or equipment supervisory monitoring. A holdup alarm may also be connected to the same channel of the transmitter as Zone C. If this is done, a holdup would be indicated as "interior alarm—day."

When the subscriber leaves, an interior alarm will be tripped (which will be reported) and the perimeter door circuit will be tripped (which will be reported). A restore signal will be reported only when all three zones have returned to normal.

The subscriber control station consists of an ace key switch and two LED's mounted in a stainless steel plate. One LED indicates that all three zones are *good.* When the subscriber turns the key to arm the system, the second LED comes on, indicating that the transmitter is reporting the closing signal. This LED goes out when the signal is received at the central station.

Because the transmitter can report multiple signals on the same call, it is practical to locate the subscriber control station near the exit door. Thus, if the subscriber arms the system and leaves immediately, the closing (interior alarm), perimeter alarm, and restore signal will all be transmitted on a single call. This will take about 30 seconds (including dialing time). The receiver at the central station will be tied up for about 15 seconds after it answers the call.

Surveillance cameras are being used extensively in banks and stores to prevent holdups, pilferage, and burglaries. Since thieves are notoriously camera-shy, the presence of a surveillance camera is often sufficient to make a would-be robber change his mind. If a business should be robbed, a surveillance camera provides sharp evidence to aid police and court.

Most surveillance cameras can be adjusted to take individual still pictures at preset intervals to keep a continuous eye on the premises. The Super 8 Kodak Surveillance camera, for example, provides continual recorded surveillance for up to 180 hours with each 100 foot roll of film. There are up to 7200 individual photographs to assist in positive identification and apprehension. This camera can be activated by a switch on the camera, by remote control, or automatically by a relay from the external alarm system.

INDUSTRIAL EQUIPMENT

Industrial security/fire-alarm systems are essentially the same as those used for commercial applications. There are, however, a few additional systems that are used more in industry than elsewhere.

Vibration detectors are often used on industrial buildings to detect vibrations caused by forced entry. Such detectors have been used on a variety of construction materials such as hollow tile, plaster and lath, brick, concrete, metal ceilings, and wood. Once mounted in place, they may be adjusted with a set screw for the desired sensitivity.

Some factories maintain a security fence equipped with fence-guard detectors. This type of detector will detect climbing, cutting, or any other penetration of the fenced area. Most of these detectors operate on standard closed-circuit controls as described previously.

Fence-guard detectors use a vertical-motion detector that is sensitive to movement created by climbing or cutting the fence. Normal side motions such as wind or accidental bumping do not affect the detector and cause false alarms. They are normally mounted about midway up the fence, and every 10 ft (3 m) of fence length. Most of these devices set off the alarm if they are tampered with or if the wire is cut. They may be connected to a control panel and the alarm will "sound" in the form of a bell or horn, or it will silently dial the local law-enforcement agency.

The only other type of detector encountered in this particular survey was the outdoor microwave detector which was used for protecting large outdoor areas like car lots, construction sites, and factory perimeters. In operation, a solid circular

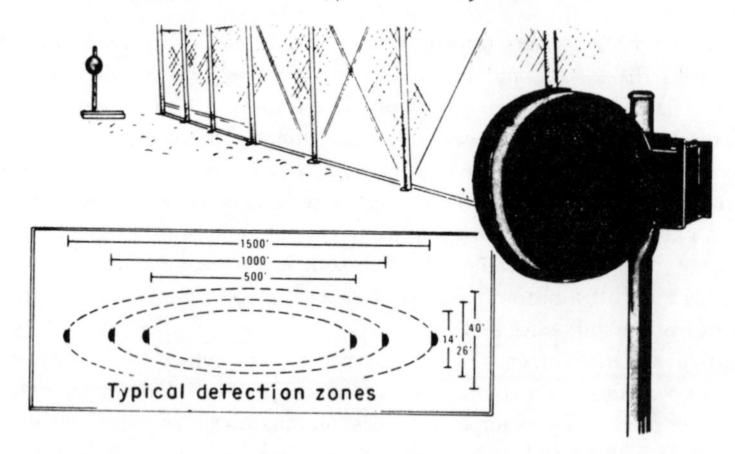

FIGURE 4-17. A solid circular beam of microwave energy extends from a transmitter to the receiver over a range of up to 1500 ft (457m).

beam of microwave energy extends from a transmitter to the receiver over a range of up to 1500 ft (457 m) (for some brands). Any movement inside of this beam (see Fig. 4-17) will activate the alarm.

THERMISTOR SENSOR

The continuous linear thermal sensor is a small-diameter coaxial wire which is capable of sensing temperature changes along its entire length. The sensor is made up of a center conductor and an outer stainless steel sheath. The center conductor is electrically insulated from the outer sheath by a ceramic thermistor material as shown in Fig. 4-18.

Since the thermistor has a negative coefficient of resistance, the electrical resistance between the center wire and the outer sheath decreases exponentially as the surrounding temperature increases (see Fig. 4-19).

The changing resistance is monitored by one of several control panels which then can actuate extinguishing systems or any other electrically controlled devices.

Such sensors have a diameter of approximately 0.080 in (0.2 cm) and therefore have a small mass which permits them to sense changes in temperature rapidly. They can sense temperatures from 70°F (21°C) up to 1200°F (649°C), if the thermistor material is properly selected.

Since electrical resistance is measured across two wires (center and sheath), the sensor has the ability to detect a high temperature on a short wire as well as a lower temperature on a longer one.

The elements are mounted by clamps spaced along their lengths and the detec-

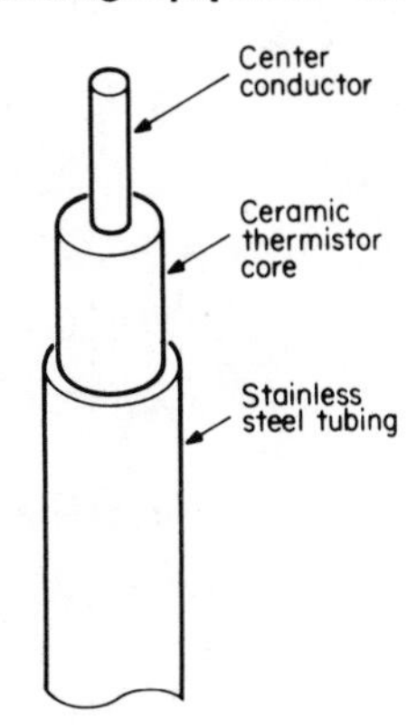

FIGURE 4-18. Structure of heat-sensor cable.

tors, being all solid state, have only two electrical failure modes: open-circuit and short-circuit. Both of these conditions can be caused only by mechanical means and are minimized by rigid mounting. Figure 4-20 shows the construction and mounting details.

ULTRAVIOLET-RADIATION FIRE DETECTORS

Ultraviolet-radiation fire detectors combine large-scale integration circuit techniques with an ultraviolet detection assembly to form a simple, yet flexible, fire-detection system.

The basis of this type of system is a gas-detection tube employing the Geiger-Mueller principle to detect radiation wave lengths extending from 2000 to 2450 angstroms (Å) (1 Å = 10^{-8} cm). Figure 4-21 displays the tube's radiation sen-

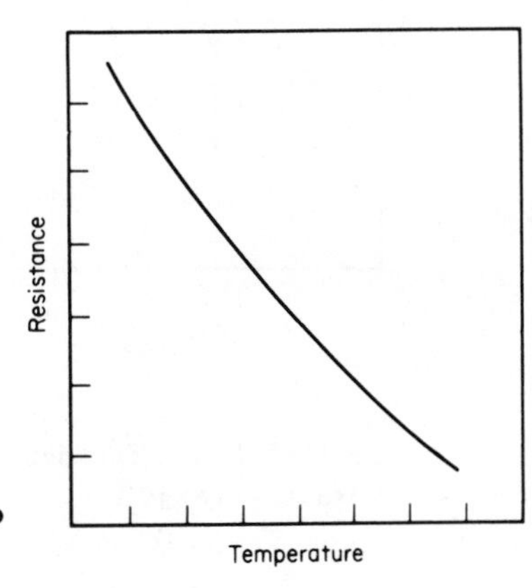

FIGURE 4-19. Curve showing relationship of resistance to temperature.

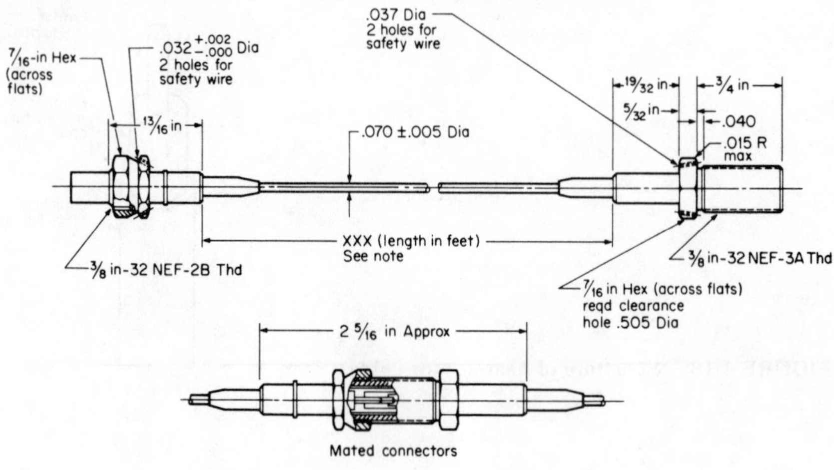

FIGURE 4-20. Using connectors to supply desired length of sensor cable.

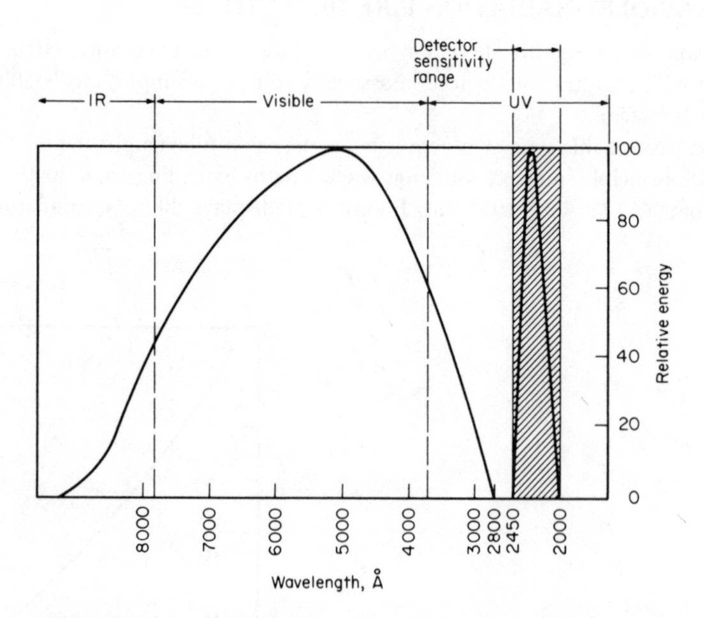

FIGURE 4-21. This detector has maximum sensitivity in the ultraviolet range.

sitive area and compares this area to other forms of radiation. It should be noted that visible radiation does not extend into the detector's sensitive area. Similarly, radiation from artificial lighting sources does not extend into the detector's sensitive area.

Welding arcs and lightning strikes, however, will generate radiation to which the detectors are sensitive and precautions must be taken to minimize these effects.

The ultraviolet-radiation detector's focus of sensitive points is a 60-degree spherical cone whose apex lies at the detector tube. Figure 4-22 indicates the relationship between viewing angle and relative sensitivity. The sensitivity of the detector tube is a characteristic of its cathode material and is fixed, but its voltage-pulse output rate varies both with flame size and flame viewing distance. The pulse output rate is directly proportional to flame size; that is, it increases when larger flame fronts are presented to the detector. The pulse output rate is also inversely proportional to the distance of the flame front from the detector tube—the pulse output rate decreases as the distance from the detector tube to the flame front increases.

To illustrate, a 1 ft^2 (0.09 m^2) hydrocarbon fire will cause a pulse output rate of 3 pulses per second at a viewing distance of 30 ft (8 m). This same fire will

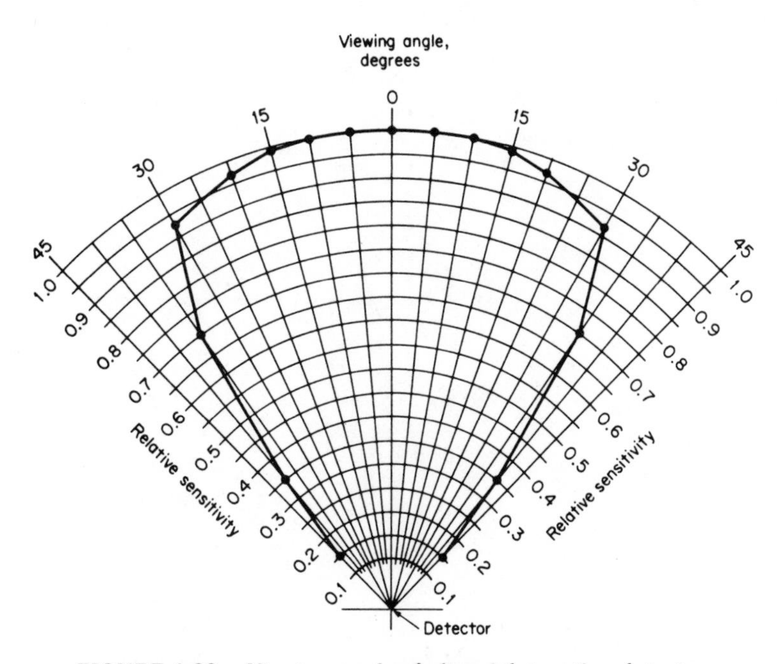

FIGURE 4-22. Viewing angle of ultraviolet motion detector.

cause a tube pulse output rate of 20 pulses per second at a viewing distance of 20 ft (6 m). In a like manner, 1 ft^2 (0.09 m^2) flame front must be located at a distance of 5 ft (1.5 m) to create a pulse output rate of 30 pulses per second, a 16 ft^2 (1.4 m^2) fire will create the same pulse output rate at a distance of 25 ft (7.6 m), and so forth.

5

DESIGN OF RESIDENTIAL SECURITY/FIRE-ALARM SYSTEMS

A wide variety of alarm systems and accessories are available to take care of almost every conceivable residental application. For example, the pictorial diagram in Fig. 5-1 shows several accessories available for residential application. Figure 5-2 shows these same accessories in an actual installation.

The residential floor plan in Fig. 5-3 shows the overall layout of a typical home with its alarm equipment as indicated by the numerals 1 through 7. Typical equipment is pictured in Fig. 5-4 and a description of each follows.

Recessed magnetic contacts (no. 3) are used at the front and side doors. A detail of their installation is shown in Fig. 5-5. The contacts should be located as far away from the hinged side of the door as practical. Holes are drilled in the doors and casings—directly across from each other—and then a pair of wires from the positive side of the protective circuit is run out through the switch hole as shown in Fig. 5-5. The switch and magnet are then installed with no more than a ⅛-in (0.3-cm) gap between the flanges of each.

All double-hung windows utilize a different type of magnetic contact. These are designed so that as long as the switch and magnet are parallel and in close proximity when the window is shut, they may be oriented side-to-side, top-to-side, or top-to-top. A pair of wires is required at each contact location, tying only the positive leg into the switch; the negative leg should not be broken.

The casement window in the kitchen is protected with recessed magnetic contacts (no. 3)—just like those used on the front and side doors.

BEGINNING THE DESIGN

The first decision to be made and the hardest to determine is what the system must accomplish, that is, identify the threat—whether it is personal risk encoun-

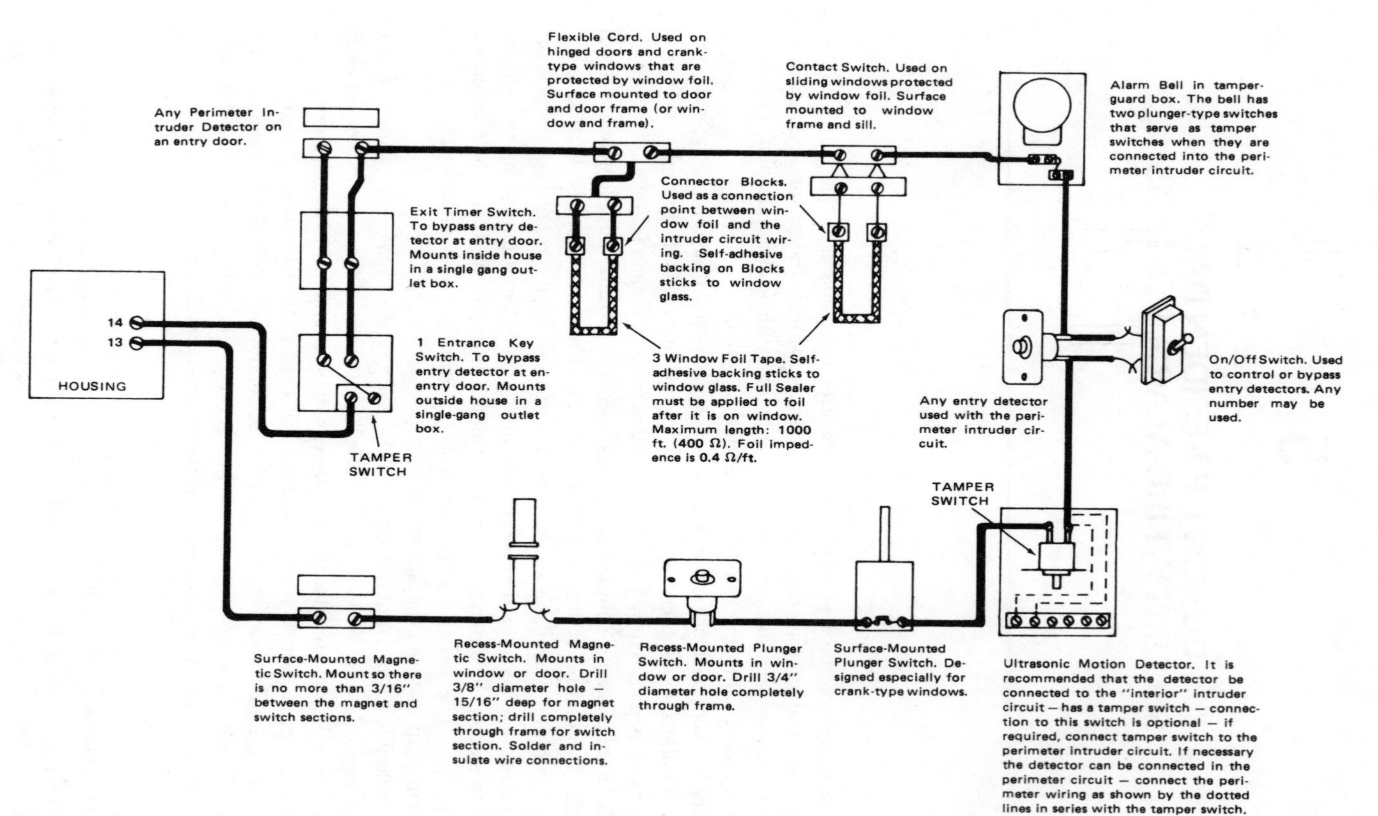

Any Perimeter Intruder Detector on an entry door.
Flexible Cord. Used on hinged doors and crank-type windows that are protected by window foil. Surface mounted to door and door frame (or window and frame).
Contact Switch. Used on sliding windows protected by window foil. Surface mounted to window frame and sill.
Alarm Bell in tamper-guard box. The bell has two plunger-type switches that serve as tamper switches when they are connected into the perimeter intruder circuit.
Exit Timer Switch. To bypass entry detector at entry door. Mounts inside house in a single gang outlet box.
Connector Blocks. Used as a connection point between window foil and the intruder circuit wiring. Self-adhesive backing on Blocks sticks to window glass.
14
13
HOUSING
1 Entrance Key Switch. To bypass entry detector at entry door. Mounts outside house in a single-gang outlet box.
3 Window Foil Tape. Self-adhesive backing sticks to window glass. Full Sealer must be applied to foil after it is on window. Maximum length: 1000 ft. (400 Ω). Foil impedence is 0.4 Ω/ft.
On/Off Switch. Used to control or bypass entry detectors. Any number may be used.
Any entry detector used with the perimeter intruder circuit.
TAMPER SWITCH
TAMPER SWITCH
Surface-Mounted Magnetic Switch. Mount so there is no more than 3/16" between the magnet and switch sections.
Recess-Mounted Magnetic Switch. Mounts in window or door. Drill 3/8" diameter hole — 15/16" deep for magnet section; drill completely through frame for switch section. Solder and insulate wire connections.
Recess-Mounted Plunger Switch. Mounts in window or door. Drill 3/4" diameter hole completely through frame.
Surface-Mounted Plunger Switch. Designed especially for crank-type windows.
Ultrasonic Motion Detector. It is recommended that the detector be connected to the "interior" intruder circuit — has a tamper switch — connection to this switch is optional — if required, connect tamper switch to the perimeter intruder circuit. If necessary the detector can be connected in the perimeter circuit — connect the perimeter wiring as shown by the dotted lines in series with the tamper switch.

FIGURE 5-1. Several security/fire-alarm accessories available for residential applications. (See Fig. 4-1.)

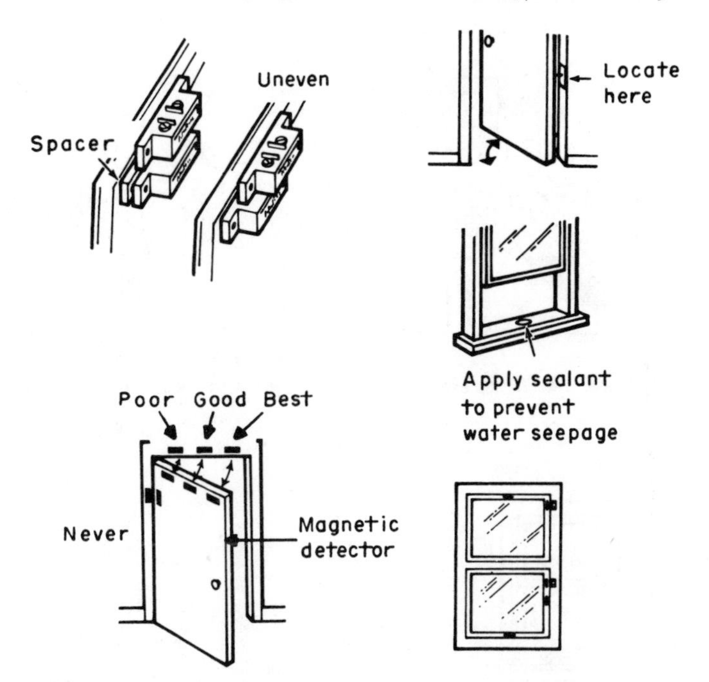

FIGURE 5-2. Practical application of the accessories shown in Fig. 5-1.

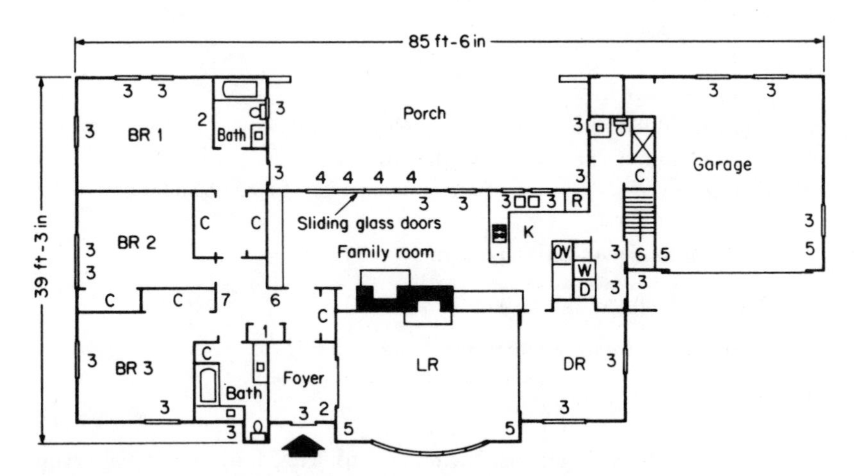

FIGURE 5-3. Residential floor plan showing the overall layout of a security/fire-alarm system. Key: 1 = Control panel; 2 = remote station; 3 = magnetic contacts; 4 = glass-break detectors; 5 = photoelectric detectors; 6 = smoke detector; 7 = fire horn.

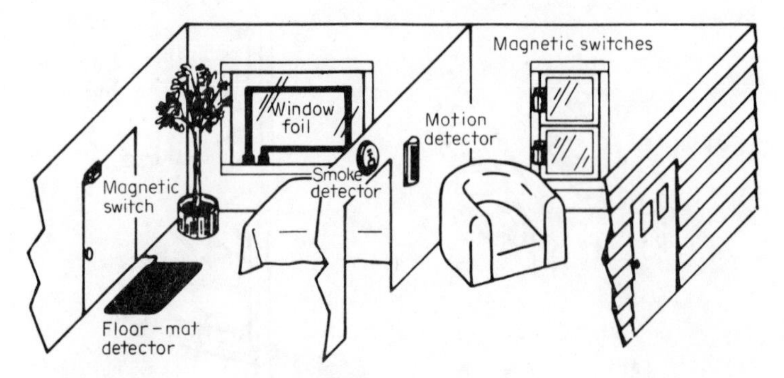

FIGURE 5-4. Placement of typical alarm system components.

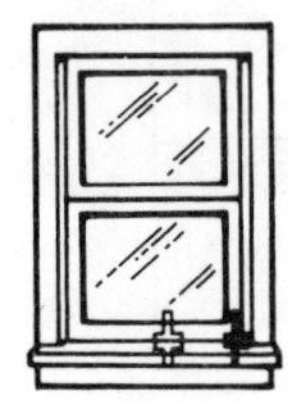

FIGURE 5-5. Installation detail of recessed magnetic contacts.

tered in a home due to burglary or psychological harassment or loss of property and money due to burglary, or some other threat.

Furthermore, the designer must decide what type of alarm should be provided.

1. Scare an intruder with an alarm sounder?
2. Call for help from neighbors?
3. Summon police or others over telephone lines?
4. Combination of the above?

The allotted budget is another factor to consider. Many residential security fire-alarm systems can be installed for less than $1000 while others may run into the thousands of dollars. The best way to determine the amount of protection needed is to consider the threat of personal risk or bodily injury and the possible loss of property. Then determine how much this protection is worth to the homeowner. When this figure is determined, you will have a good guide to how much of an alarm system is needed.

With the threat determined, the system goals and a general budget figure for the system cost established, the design may begin.

BASIC INFORMATION

A form such as the one in Fig. 5-6 should be used by the contractor or installer to ensure that all necessary information concerning the design will be obtained. Contractors should have a supply of such forms printed with their letterhead on top of the page. Manufacturers of security/fire-alarm equipment also furnish such forms at little cost. You will also need the information included in the form in Fig. 5-7.

With the information shown in Figs. 5-6 and 5-7, the designer should now begin to think like a burglar and examine the premises from this viewpoint. List the most likely points of entry and determine why.

Burglars almost always prefer doors for entrance since doors are usually required for removing such items as television sets from the home. Furthermore, they need a concealed approach route and a hiding place for the vehicle to be loaded with the stolen goods. Therefore, back alleys and hidden carports are the obvious vehicle hiding places. With this knowledge in mind, choose the most vulnerable door in the building as deserving the best physical and burglar-alarm protection. A dead latch and dead-bolt locks should be used on a solid wood or steel

BASIC INFORMATION

Name ______________________ Date of Survey ____________

Street ____________________ City ______________ Zip ______

Telephone ____

Family size: Adults __________ Ages ____________________

Children __________ Ages ____________________

Pets __________ Weight ____________________

Normal bedtime ____________ How many in family smoke? ____________

Name of closest relative ______________ Telephone ____________

Name of insurance agent __________ Company __________ Tel. ______

FIGURE 5-6. Form used for initial survey of project.

CONSTRUCTION INFORMATION

Age of home: ______ Lot size: ______ Ft. wide ______ Ft. deep ______

Acre(s) ______

Building measurements: ______ Ft. long ______ Ft. wide ______ Ft. high ______

Walls: Wood ____ Shingle ____ Stone ____ Brick ____ Metal siding ____

Other ______________________

Interior walls: Wood ____ Plaster ____ Plasterboard ____

Flagstone ____ Brick ____ Fiberboard ____

Other ______________________

Ceilings: Wood ____ Plaster ____ Plasterboard ____ False ____ Acoustic ____

Other ________

Roof: Wood shingle ____ Tar paper ____ Tile ____ Asphalt shingle ____

Rock ____ Slate ____ Pitched ____ Flat ____ Multistory ____

Other ______________________

Floors: Wood ____ Concrete ____ Tile ____ Flat stone ____ Carpet ____

Other ________

Attic: Full ____ Partial ____ None ____

Attic crawl space: Good ____ Average ____ Poor ____ None ____

Crawl space under house: Good ____ Average ____ Poor ____ None ____

Basement: Full ____ Partial ____ None ____ Basement dimensions: ______ x ______

Garage: Attached ____ Free standing ____ Garage dimensions: ______ x ______

Garage walls: Wood ____ Unfinished ____ Plasterboard ____ Brick ____

Stone ____ Other ______________________

Type of insulation:

Walls: ________ Floor: ________ Ceiling: ________ Basement: ________

FIGURE 5-6. Form used for initial survey of project. (*Continued*)

INTRUSION SECURITY INFORMATION

Law-enforcement agency ____________ Telephone ________ Response time ______

Distance to neighbor on:

Right __________ ft. Left __________ ft. Front __________ ft. Rear __________ ft.

Type of public lighting: Street _____ Flood _____ Other __________________ None _____

Type of private lighting: Front _____ Side _____ Rear _____

What lights are left on at night? ______________________________

Visibility onto property from street:

Good _____ Average _____ Poor _____ Why? ______________________________

Hours home is normally vacant: ______________________________

Number of entrances:

Doors __________ Sliding glass doors __________ Windows __________ Other ________

Doors: Wood _____ Metal _____ Glass _____ Single _____ Double _____ Hollow _____

Solid _____

Door frames: Wood _____ Metal _____ Aluminum _____ Other ____________________

Locks: Single-key _____ Sliding-bolt _____ Double-key ____ Dead-bolt _____ Padlock _____

Night latch _____ Other __________________ When last rekeyed? __________

Windows: Sliding _____ Double-hung _____ Single-hung _____ Casement _____

Louver _____ Fixed _____ Other ______________________________

Glass: Single-strength _____ Double-strength _____ Plate _____ Frosted _____

Tempered _____ Thermal pane _____ Other ______________________________

Storm windows? Yes ☐ No ☐

Glass framing: Wood _____ Metal _____ Other ______________________________

Grilles or Screening: Standard screens _____ Iron bars _____ Other ____________________

Garage door: Single _____ Double _____ One-piece overhead _____

Sectional overhead _____ Swing out _____ Other ______________________________

Outbuildings: Number _____ Type ______________________________

FIGURE 5-7. Form to obtain intrusion security information.

steel door with no windows in it, if possible. If this point of entry is a sliding glass door, care must be taken so that the door cannot be lifted out of the track and the door should be capable of being securely key-locked and pinned into the closed position using appropriate hardware.

Obviously, this most vulnerable entry point must be alarmed with the best equipment available. Similarly, other doors of the home must have some sort of protection.

Windows should be analyzed in a similar manner to doors. In most premises there are one or more windows which may face a side yard, a fence, or a wall where no casual observer is likely to see someone making forced entry. First, protect the window from being forced open and then install a dependable security system.

Most security systems use a closed-loop protective circuit where a pair of wires is connected to the alarm control and is then run around the perimeter of a building and finally returned to the alarm control panel. Closed-circuit detectors are connected in series in this loop. A small current flows through the wiring and detectors and any interruption of this current by the detector operation (cutting the wires or shorting the wire pair together) will sound the alarm. Restoring the loop to its original condition, such as closing the alarm door after entry, will not stop the alarm condition. Only operating the appropriate control will do this.

Magnetic contacts or switches are by far the most commonly used detection devices for openings such as windows and doors. They consist of two pieces—a magnet and a magnetically operated switch enclosed in plastic cases. The magnet is mounted on the edge of the door while the switch section is mounted directly adjacent to the magnet on the door frame. When the magnet is located near the switch section, the switch is turned on and electricity flows through the switch contact. Moving the magnet away from the switch, such as opening a door, turns the switch off.

Since this is a closed-circuit system, the current through the loop will cease and the alarm will sound on opening. Magnetic switches are very successful because they are noncritical in alignment between the magnet and switch section and are extremely reliable in operation. Many switches are rated for hundreds of millions or even billions of switch operations. There is little mechanical motion in this switch, so replacement will be extremely infrequent under the worst of circumstances. By their nature, they are free from false alarms and are easy to troubleshoot and replace in the event one fails.

Magnetic contacts are also the best method of protecting windows and other openings. To protect glass from breakage in windows or sliding glass doors, a special lead foil is the common means of protection. This foil is put in series with the same burglar circuit that connects the doors and windows. The alloy in the foil is of such composition that any break in the glass will break the foil and thereby set off the alarm.

It may also be advisable to include an extra door switch or two on some of the interior doors that are likely to be opened in the event an intruder somehow penetrates the perimeter circuit. Such doors might include those to a gun closet, fur storage vault, or just between two rooms that have to be traversed to find any valuable property. The intruder is likely to have his guard down at this point and not to be looking for such a switch. Motion detectors—such as ultrasonic, infrared, audio, etc.—are also good insurance for the interior circuit.

Routing the circuit wires around the perimeter in an effective manner is one of the most important parts of a security/fire-alarm system. A pair of either 22-AWG or 24-AWG wires should be run all the way around the home from the control panel and then back. All detectors are then connected to this perimeter loop.

Wire concealment can be a major problem for the installer. If the house is under construction, the pair of wires can be located at some set distance within the partitions and walls. The installer can then cut into the wall at this distance when the walls are finished to get to the wires for the final connections. For existing construction, much fishing is necessary to route the wires to the various detectors, but Chap. 3 gives several solutions to these problems.

Concealment is important for aesthetic reasons, for making it impossible for the intruder to locate the presence of the system, and for reliability in the sense of minimizing damage to the wires.

RESIDENTIAL FIRE-ALARM SYSTEM

Heat and smoke detectors should be included in any residential security/fire-alarm system. They are generally connected to the system as shown in Fig. 4-1. The fire-detection circuit should be fully supervised as required by UL (Underwriters Laboratory). The circuit itself should act as a detector in the event of a malfunction; that is, a trouble bell or buzzer should activate in the control unit to alert the occupants of the situation.

The primary location for installing smoke detectors is outside each bedroom area. Since fire travels upward, the top of each stairwell is another important location. The National Fire Protection Association (NFPA) also recommends that smoke detectors be installed on each living level of a multistory house.

Heat detectors should be installed in each enclosed living area including bathrooms, closets, attics, and basements. Any number of detectors can be used with most fire-alarm systems.

Rate-of-rise heat detectors should be mounted on the ceiling not less than 6 in (15 cm) from a side wall. Exact detector location can be determined by an evaluation based on good engineering judgment, supplemented if possible by field tests.

The chart in Fig. 5-8 shows some of the heat/smoke detectors supplied by NuTone. The model number, a description of each component, suggested use, and dimensions are given to assist designers of security/fire-alarm systems.

For further information on the design of residential fire alarm systems, contact the National Fire Protection Association, 470 Atlantic Avenue, Boston, MA

HEAT-SMOKE DETECTOR SELECTION GUIDE
(Use with NuTone SA-2300 or S-2100 Security Alarm Systems)

Model no.	*Description*	*Suggested use*	*Specifications*
S-120	135°F fixed-temperature heat detector	Surface-mount on ceiling in ordinary living areas with normal room temperatures.	1¾ in diameter, ¾ deep Distance range: 10 ft in all directions. Detector covers area up to 20 x 20 ft.
S-121	200°F fixed-temperature heat detector	Surface-mount on ceiling in areas where temperatures are higher than normal: furnace or boiler rooms; attics.	
S-122	Rate-of-rise/135°F fixed-temperature heat detector	Surface-mount on ceiling in ordinary living areas with normal room temperatures.	4½ in diameter, 1⅜ in deep Distance range: 25 ft in all directions. Detector covers area up to 50 x 50 ft.
S-123	Rate-of-rise/200°F fixed-temperature heat detector	Surface-mount on ceiling in areas where temperatures are higher than normal: furnace or boiler rooms. Note: Use S-121 or S-125 in areas where temperatures consistently exceed 150°F.	
SA-124	135°F fixed-temperature heat detector	Surface-mount on ceiling in ordinary living areas with normal room temperatures.	1¾ in diameter, ¾ in deep Distance range: 15 ft in all directions. Detector covers area up to 30 x 30 ft.
SA-125	200°F fixed-temperature heat detector	Surface-mount on ceiling in areas where temperatures are higher than normal: furnace or boiler rooms; attics.	
S-245	Smoke detector	Surface-mount on 4-in square or octagonal wiring box, primarily outside bedrooms.	5½ in square x 2⅝ in deep
S-245H	Smoke detector with 135°F heat sensor		
S-240	Smoke-detector supervisory module for S-2100 *only*	Installs in wiring box at location of last smoke detector in series.	2 in x 2 in x 1 in

FIGURE 5-8 Application of heat/smoke detectors

02210, and request a copy of NFPA no. 72E, Standard on Automatic Fire Detectors.

While no regularly scheduled maintenance is necessary for most heat/smoke detectors, periodic cleaning of the detection chambers may be required when detectors are located in abnormally dirty or dusty environments.

PRACTICAL APPLICATIONS

To better understand the procedures necessary to design a suitable residential security/fire-alarm system, the floor plan of a two-story residence is shown in Fig. 5-9. The obvious starting place for the design of the system is at the normal

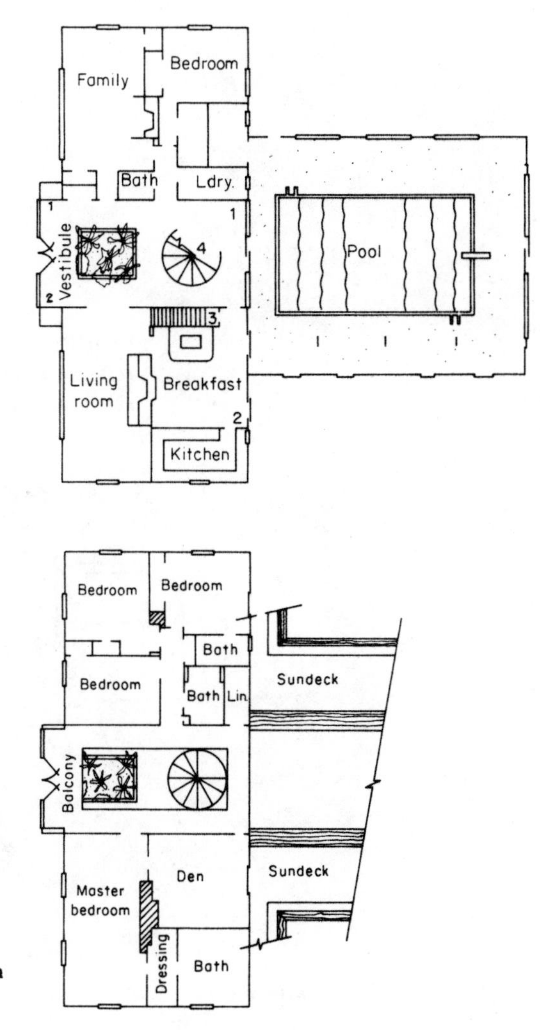

FIGURE 5-9. Floor plan of a two-story residence.

entrances, such as the front doors opening into the vestibule and the sliding glass doors in the rear of the house opening into the vestibule and breakfast/kitchen areas. These types of entries may be protected by several methods, but in this case, infrared photoelectric entry detectors seem to be the best.

For example, transceivers are positioned at the locations indicated by the numeral 1 and reflectors are located at the locations indicated by the numeral 2. Each of these items resembles a conventional quadruplex receptacle as shown in Fig. 5-10. The centers of these outlets are located approximately 18 in (45 cm) above the finished floor so that an intruder will break the beam as shown in Fig. 5-11.

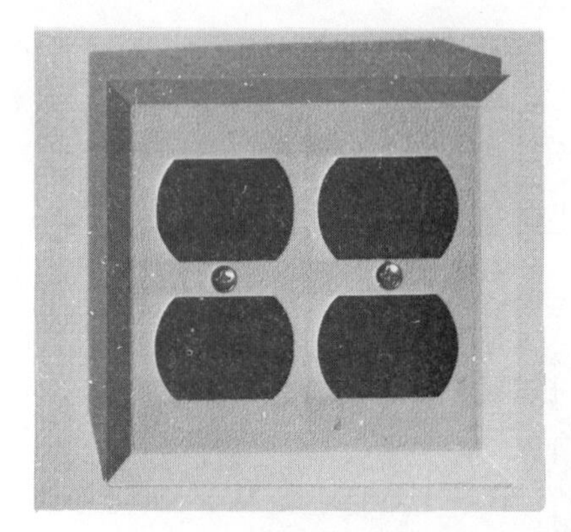

FIGURE 5-10. Transceivers and reflectors resemble ordinary quadruplex receptacles.

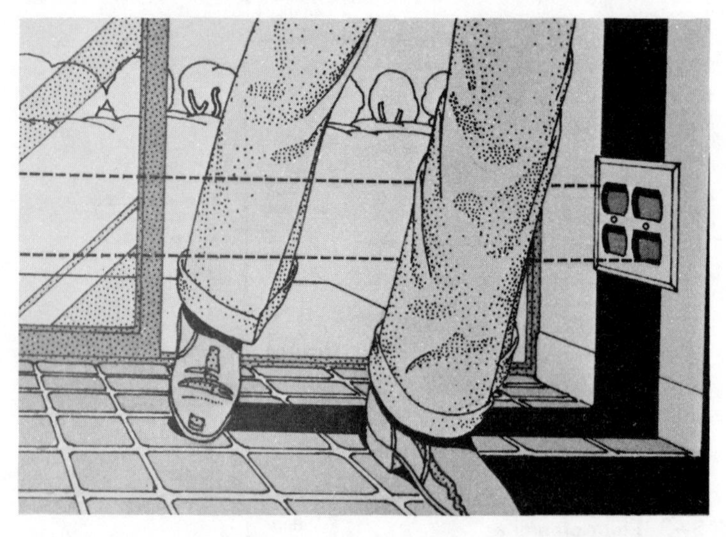

FIGURE 5-11. Transceivers and reflectors are located approximately 18 in above the floor so that an intruder will break the beam.

Note that the infrared photoelectric entry detectors guarding the rear entries protect three separate doors since the transceiver is located in the vestibule and aimed at a reflector mounted in the end of a kitchen cabinet.

This distance may seem great, but as long as there are no obstacles between the transceiver and reflector (furniture, plants, etc.) and the distance is no more than 75 ft (23 m) apart, the system will function properly.

At the top of the basement stairs, a floor-mat entry detector (indicated by the numeral 3) is positioned. The mat is shown in Fig. 5-12 and should be concealed by a scatter rug, as shown in Fig. 5-13. Floor-mat detectors should also be located on stairways (numeral 4), as shown in Fig. 5-14. They may also be located in other interior locations that are likely to be used by intruders.

The items described thus far may be termed interior protection. Now, perimeter protection must be provided to ensure an intruder-proof home. Window foil

FIGURE 5-12. Typical floor-mat entry detector.

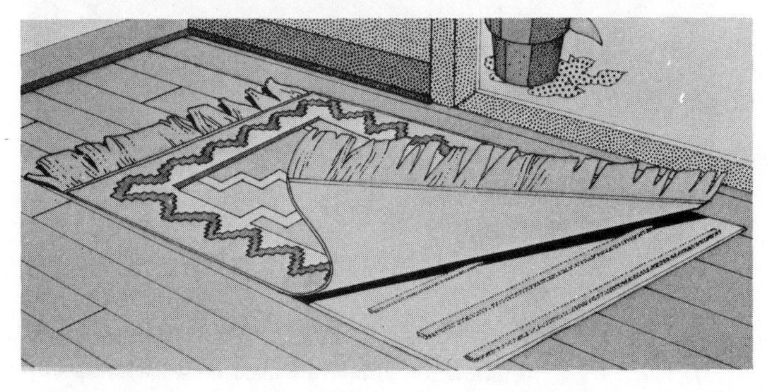

FIGURE 5-13. A scatter rug is a good medium for concealing floor-mat entry detectors.

FIGURE 5-14. Floor-mat detectors may also be located on stairways.

(Fig. 5-15) should be used on all windows and possibly the rear sliding glass doors. Doors may be protected by recessed magnetic detectors (Fig. 5-16) or recessed plunger detectors, as shown in Fig. 5-17. Figure 5-18 shows some possible locations for this type of detector.

Of course, the system will need a delayed-entry control as shown in Fig. 5-19

FIGURE 5-15. Window foil should be used on all windows as well as on sliding glass doors.

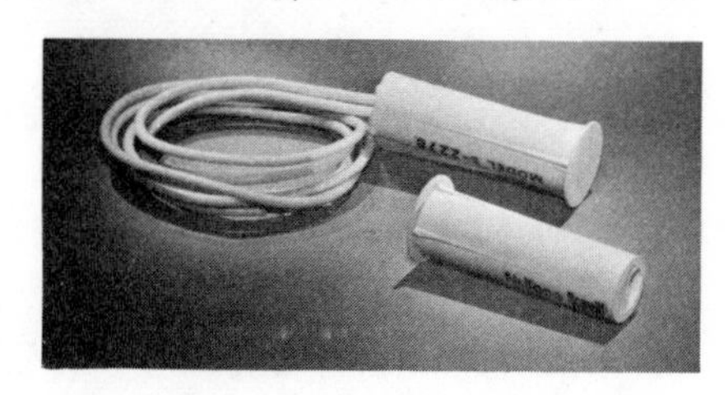

FIGURE 5-16. Doors may be protected by recessed magnetic detectors.

and some means of sounding an alarm. A bell (Fig. 5-20), horn (Fig. 5-21), or telephone dialer (Fig. 5-22) may be used. The wiring of all of these units is performed as discussed previously.

If the security/fire-alarm system is operated by conventional house current, you should have a battery backup system. Also, the designer should consider some possible causes of false or unwanted alarms, as presented in the following list:

1. Severe electrical storms.
2. Faulty smoke detector.
3. Faulty wiring: wire connections, staple cutting through insulation, insulation broken by severe bending, closely spaced bare wires which may touch if jarred by vibration of refrigerator, washer, dryer, furnace, etc.
4. Electrical transients from heavy-duty appliances, such as refrigerators, relays, etc.

FIGURE 5-17. Recessed plunger detectors may also be used for door protection.

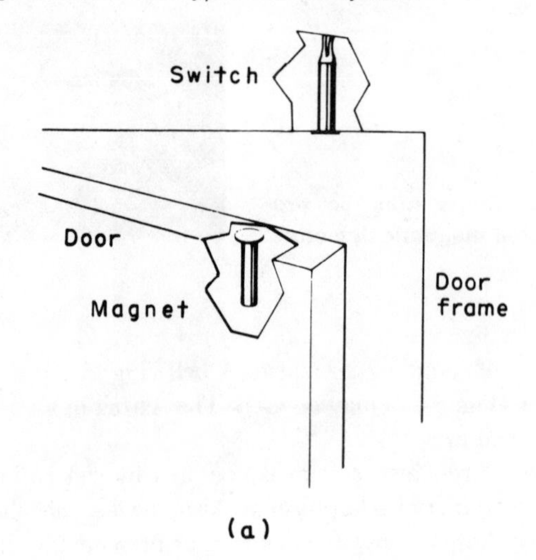

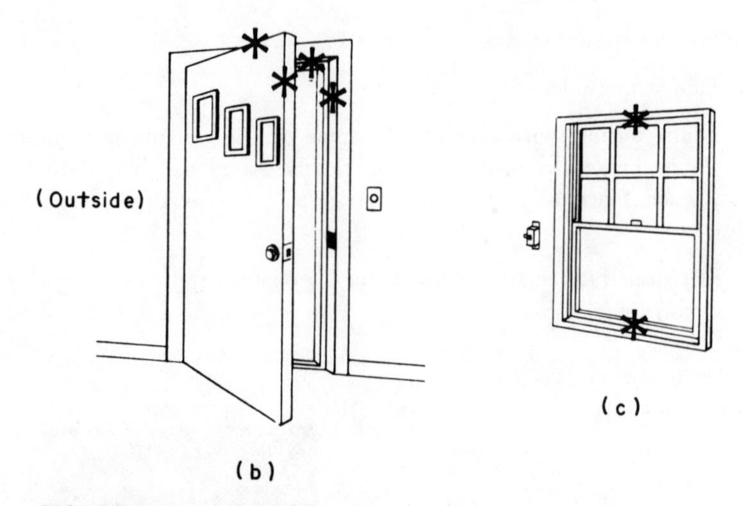

FIGURE 5-18. Some of the possible locations for recessed magnetic entry detectors.

FIGURE 5-19. A key-operated delayed entry control.

FIGURE 5-20. A bell may be used as a means of sounding the alarm.

FIGURE 5-21. A horn is often used to alert neighbors that an intruder is on the premises.

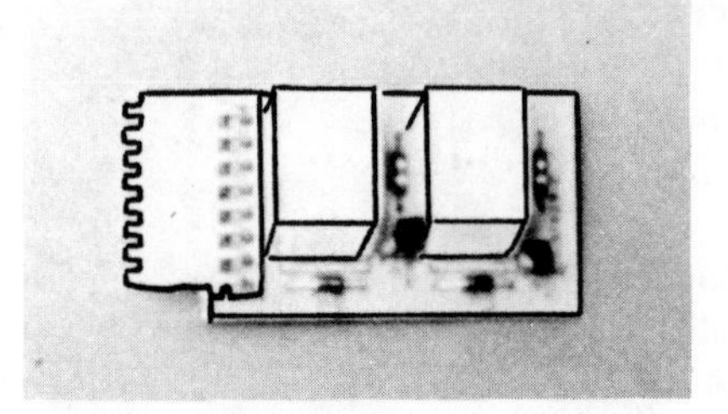

FIGURE 5-22. A telephone dialer is one means of notifying law-enforcement agencies without scaring off the intruder.

5. The use of low-temperature heat detectors in a high-temperature environment, such as attic and furnace room.

6. Concentration of sunlight on a heat detector or smoke detector.

7. Accidental activation of an intruder detector (opening protected door or window, exerting 70 lb or more of pressure on floor mat, depressing an emergency alarm push button).

8. A momentary activation of an entry detector switch on the perimeter or interior detection circuit, caused by a severe vibration.

9. Shortwave or C.B. radio operating with excessive power near your home.

Standpipes, fire hoses, and sprinkler systems are required in many commercial buildings. In many states, the regulations are controlled by the state fire marshall.

In general, a Class II service fire standpipe requires a minimum residual pressure of not less than 65 psi at the topmost outlet. Standpipes should be located in noncombustible fire-rated stair enclosures; in multistory buildings exceeding 275 ft in height, the fire protection standpipe system must be zoned accordingly.

Fire standpipe risers must be designed so that a stream of water can be brought to bear on all parts of all floors within 30 ft of a nozzle. The nozzle, in turn, is connected to not more than 100 ft of riser-attached hose.

For Class II service, fire standpipes shall be provided with 1½-in hose connections on each floor. However, a fire standpipe in excess of 100 ft in height shall be a minimum of 6 in in size at its base. Furthermore, at least one fire department hose valve shall be provided at each floor level for fire department use in any building under construction.

The number of standpipes in each building will determine the minimum water flow. For example, a building with nine standpipes—conforming to Class III fire protection service—must have an automatic fire pump providing at least 2500 gpm. A 750-gpm capacity fire pump for a standpipe system may be provided with three 2½-in hose wall outlets at ground level for fire department use. Fire standpipes under 50 ft high shall be a minimum of 2 in in diameter. However, many local codes require a minimum size of 4 in in diameter for fire standpipes, so always check the local codes and ordinances before beginning the design and certainly before anyone starts construction.

In buildings with combined systems, designed to be completely sprinklered, with risers sized by hydraulic calculations, the contractor must submit complete calculations to the authority having jurisdiction.

Ordinary Hazard (Group 3) occupancies are defined as buildings that have a high quantity and/or high combustibility of contents. Such occupancies include woodworking businesses, feed mills, and the like.

Sometimes during modifications to an existing fire protection system, it becomes necessary to remove all or a portion of the system from service. When all or any portion of a standpipe system is out of service for any reason, the local fire department shall be notified.

When a standpipe system has been out of service for a number of years, before it is filled with water and restored to service it shall be tested with air at a pressure not exceeding 25 psi to determine its tightness.

6

DESIGN OF COMMERCIAL SECURITY/FIRE-ALARM SYSTEMS

The design of commercial security/fire-alarm systems is very similar to residential systems except that heavier duty equipment is normally used and the goals are somewhat different from residential demands.

The floor plan of a small commercial building is shown in Fig. 6-1. A burglar-alarm panel and a telephone dialer are located in the storage area (5). The relay-type control panel (Fig. 6-2) has one protective circuit, output for 6-Vdc alarm-sounding devices, dry contacts to actuate other deterrent or reporting devices, and a silent holdup alarm with telephone dialer to dial emergency numbers and deliver voice messages.

Glass on the front door is protected with window foil connected to foil blocks which are then connected to the protective circuit wiring in the alarm system. Door cords (Fig. 6-3) are used to provide a flexible connection from the foil blocks on the door and window to a solid contact point adjacent to the door. Flexible door cords also may be used on moving windows and money clips installed in cash drawers. The drawing in Fig. 6-4 shows the front door on the floor plan in Fig. 6-1.

The large display window (2) is again protected with foil connected to foil blocks, while the office area is protected by an ultrasonic motion detector mounted as shown in Fig. 6-5. A surface-mounted door contact is used to protect the rear door of the building (see Fig. 6-6). Legal entry is permitted by use of a key lock.

All burglar-alarm systems have three common functions: detection, control, and annunciation or alarm signaling. Most detectors incorporate switches or relays that operate because of entry, movement, pressure, infrared-beam interruption, etc. The control senses detector operation with a relay and produces an output that may operate a bell, siren, silent alarm, and so forth. The controls also frequently contain ON/OFF switches, test meters, time delays, power supplies,

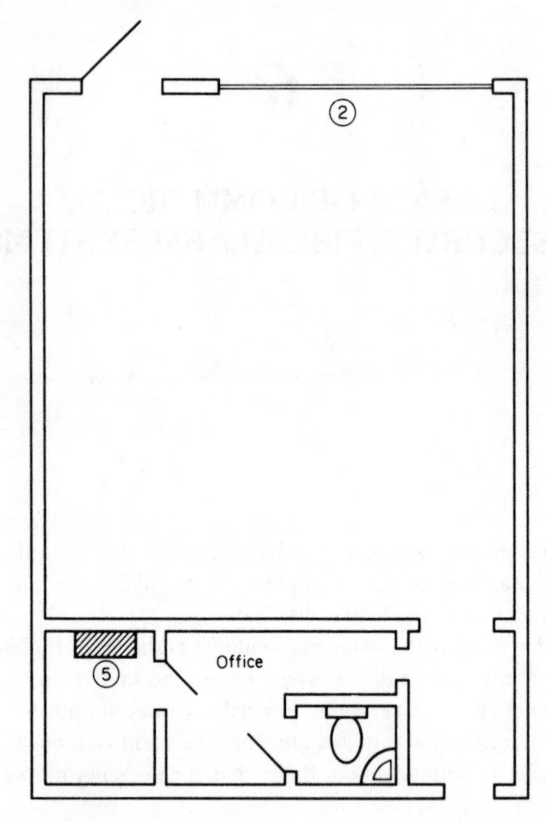

FIGURE 6-1. Floor plan of a small commercial building.

standby batteries, and terminals for tying the system together. The control output usually provides power on alarm to operate bells, sirens, etc., or switch contacts for silent alarms, such as automatic telephone dialers.

A pictorial diagram of a typical burglar and fire-alarm system is given in Fig. 6-7. This particular system can accommodate at least 20 burglar detectors, three smoke detectors, and an unlimited number of thermostat fire detectors.

The burglar-alarm circuit permits use of magnetic switches, switch mats, and ultrasonics for intruder detection. Two zones are incorporated into the unit under discussion: one zone is used for an outside perimeter guard and a second connects interior motion detectors and floor mats.

Conventional burglar-alarm systems usually employ contacts at all openings, such as doors and windows. However, the contents of some buildings are so val-

FIGURE 6-2. Relay-type control panel utilizing one protective circuit and having output for alarm-sounding devices and other detection/alarm capabilities.

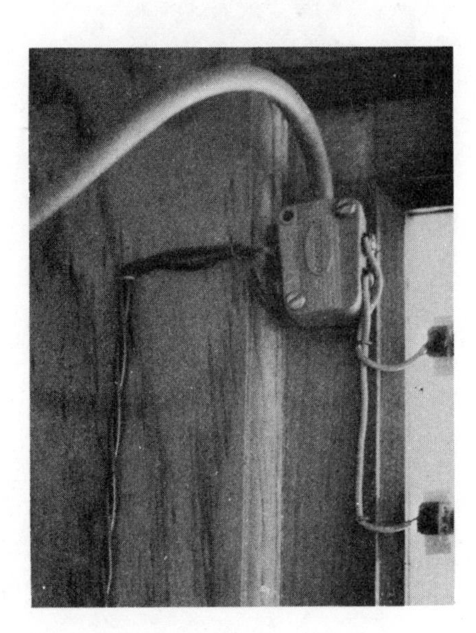

FIGURE 6-3. Door cords are used to provide a flexible connection from the foil blocks to a solid contact point.

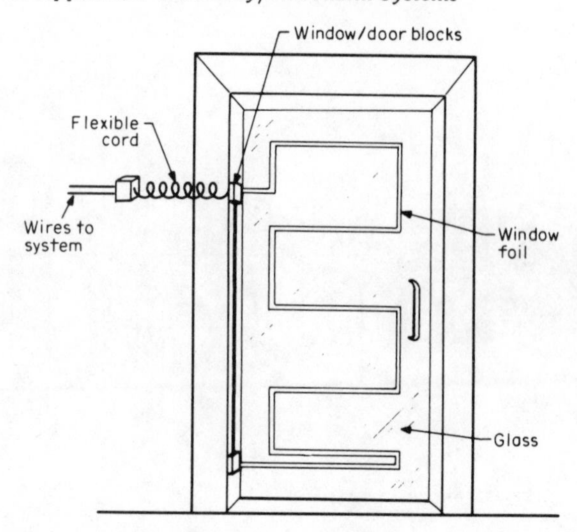

FIGURE 6-4. Elevation view of the front door on the plan in Fig. 6-1.

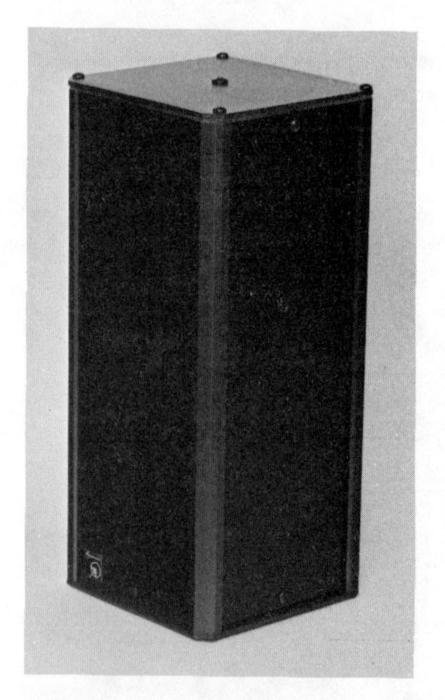

FIGURE 6-5. Ultrasonic motion detector used to protect the office area.

FIGURE 6-6. Surface-mounted door contact used to protect the rear door of the building.

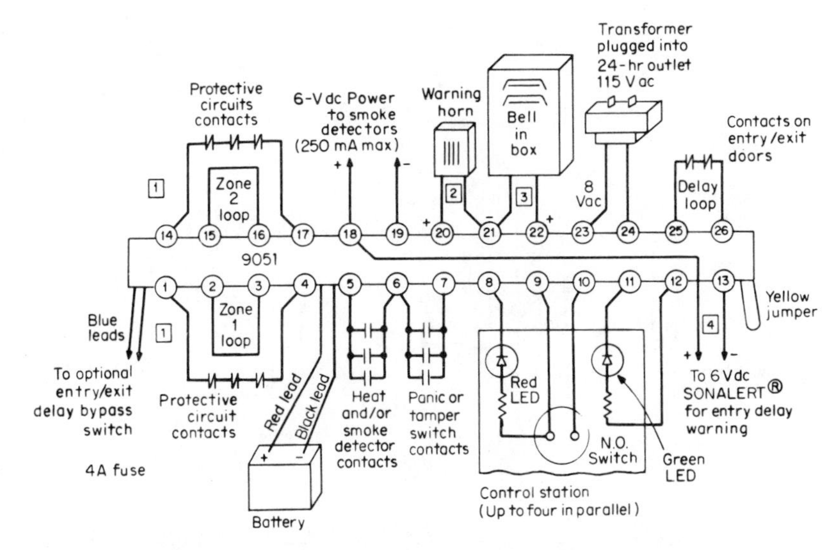

FIGURE 6-7. Diagram of a typical burglar/fire-alarm system.

uable that burglars may try to enter through walls, floors, or ceilings. The most economical method of detecting this type of forced entry is to use vibration detectors. These detectors are designed to initiate an alarm from vibration and to detect burglary attempts through walls and ceilings by impact attacks with a hammer, crow bar, or other heavy object. The detectors should be installed on brick, hollow tile, concrete, or plaster walls, about 4 ft (1.2 m) or 5 ft (1.5 m) from the floor and from 2½ ft (0.8 m) to 6 ft (1.8 m) apart, depending on wall thickness and construction. In Fig. 6-8, for example, on an 8-in (20-cm) brick wall, the detectors are from 4½ ft (1.4 m) to 5 ft (1.5 m) from the floor and on 6-ft (1.8-m) centers. Tests have shown this to be the proper spacing; however, further tests on a specific installation may indicate spacing requirements that are greater or less than those given here.

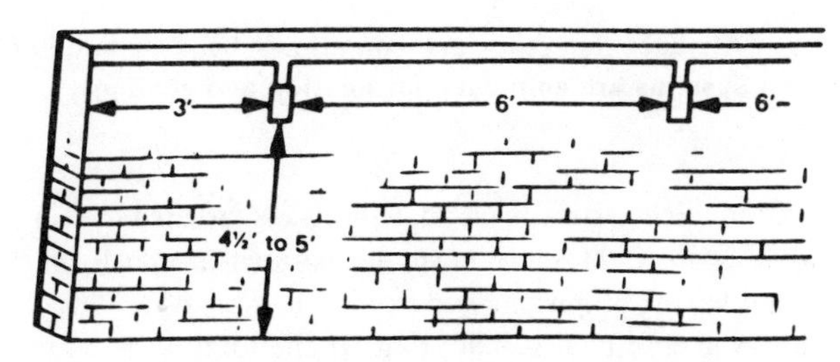

FIGURE 6-8. Vibration detectors positioned on an 8-in brick wall.

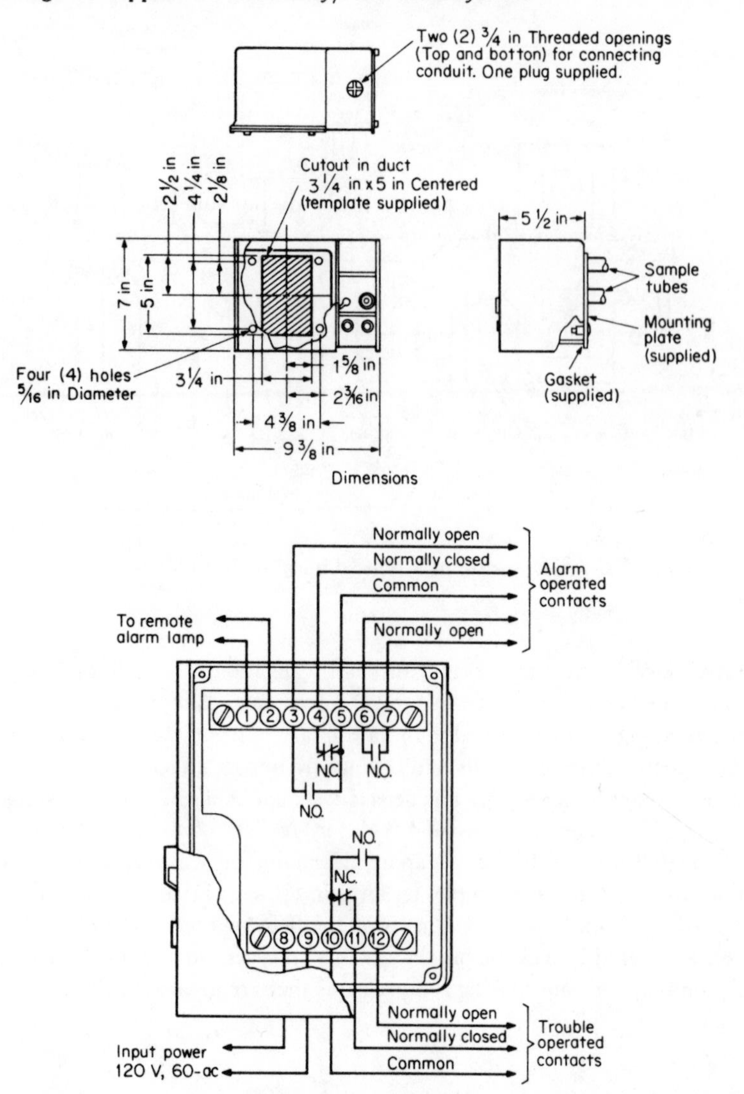

FIGURE 6-9. Air-duct detectors are sometimes used in buildings where forced-air duct systems are employed for heating and cooling.

Detectors are mounted vertically on angle brackets fastened to 1-in (2.5-cm) by 2-in (5-cm) wood strips, which are normally painted to match the area finish. Detector units and wiring of burglar and fire alarm systems are installed like any other type of low-voltage signal system; that is, one locates the outlets, furnishes a power supply, and connects the components with the proper size wire.

Some commercial systems—such as warehouses, etc.—use exposed wiring or easily concealed wires above false ceilings. These may suffice, but concealed wiring is much more secure against tampering or defect by either insiders or outsiders. Finally, appearance is always important and exposed wiring is seldom decorative.

Alarm systems using wiring for interconnection are usually superior to other types since they are more secure and will last longer. Other types of connection, such as those using radios or building wiring, are not supervised and depend on relatively unreliable electronic circuits.

Alarm systems are often classified according to the means of sounding the alarm or alerting someone who can respond effectively to a break-in or fire. A local alarm, for example, includes a detector or control unit to provide ON/OFF, testing, power, etc., and a signaling device, such as a sounder light, to indicate the alarm locally. The disadvantage of a local alarm is that the intruder knows that he has been detected and can easily avoid getting caught.

Another type of alarm is the silent alarm. This system utilizes a detector, a control, and a means of sending an alarm by wire to a remote location where action can be initiated. The alarm-control output may be sent by leased telephone line to a central station where professionals can follow up with police action. Another kind of silent alarm uses an automatic telephone dialer that utilizes phone service to contact several parties such as the police, fire department, alarm company, owner, or manager.

COMMERCIAL FIRE-ALARM SYSTEMS

The purpose of any fire-alarm system is to save lives and property. Every fire-safety authority agrees that the greatest single factor contributing to low fatalities and property loss in fire is early detection and warning. In fact, it is generally agreed that what happens in the first 3 minutes of a fire represents the difference between minor inconvenience and major catastrophe. To accomplish this vitally important early detection and warning, designers locate fire-alarm stations at frequent intervals in corridors, auditoriums, and other places that are usually occupied and automatic fire detectors in boiler rooms, kitchens, ventilating shafts, attics, storage rooms, and so on, where a hidden fire might not otherwise be detected in its early stage.

Besides conventional equipment, air-duct detectors are sometimes used in buildings where forced-air duct systems are employed for heating and cooling. These are enclosed sectional metal cabinet units containing an ionization detector and control elements. The unit is normally mounted to the external side of the duct, details of which are shown in Fig. 6-9.

Many older fire-alarm systems can be updated—utilizing modern ionization, infrared, and photoelectric detection capabilities—by installing a special control panel to the old system. Most of these control panels can be connected to most coded and noncoded systems and operate almost any number of ionization and thermal detectors up to five infrared or photoelectric detectors as well as use the existing system to sound the alarm.

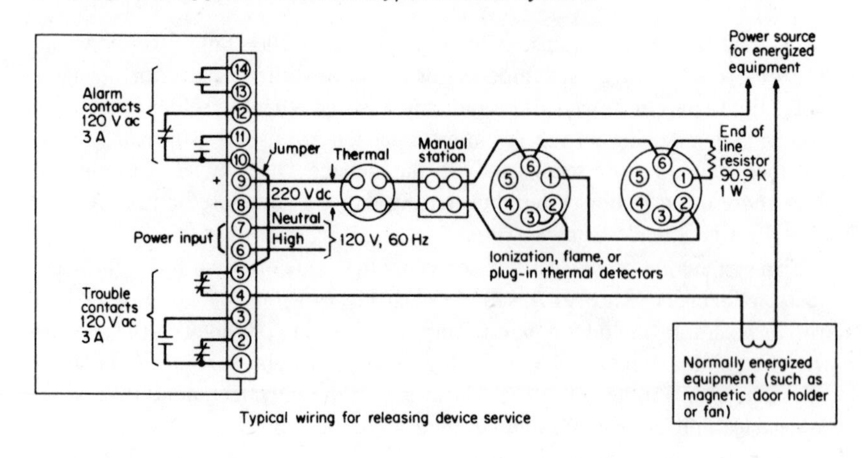

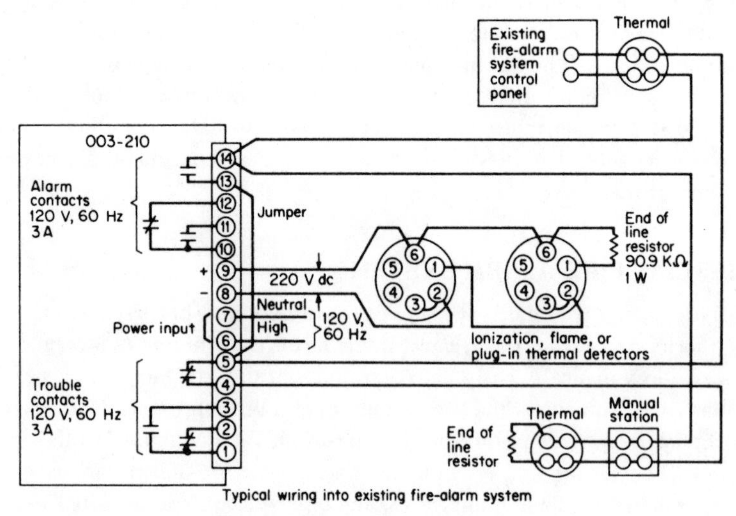

FIGURE 6-10. Typical fire-alarm system utilizing modern detecting devices.

In addition to the above, fire-alarm systems can be used to directly release smoke barriers, shut down equipment, actuate extinguishing systems, open smoke vents, and so on. A typical system is shown in Fig. 6-10.

PRACTICAL APPLICATIONS

The floor plan of a small courthouse is shown in Fig. 6-11. Note that the fire-alarm panel is located in the mechanical room on another floor and is not shown in this drawing. However, all striking stations, bells, door-release switches, and ionization smoke detectors are shown on this plan and operate in the same manner

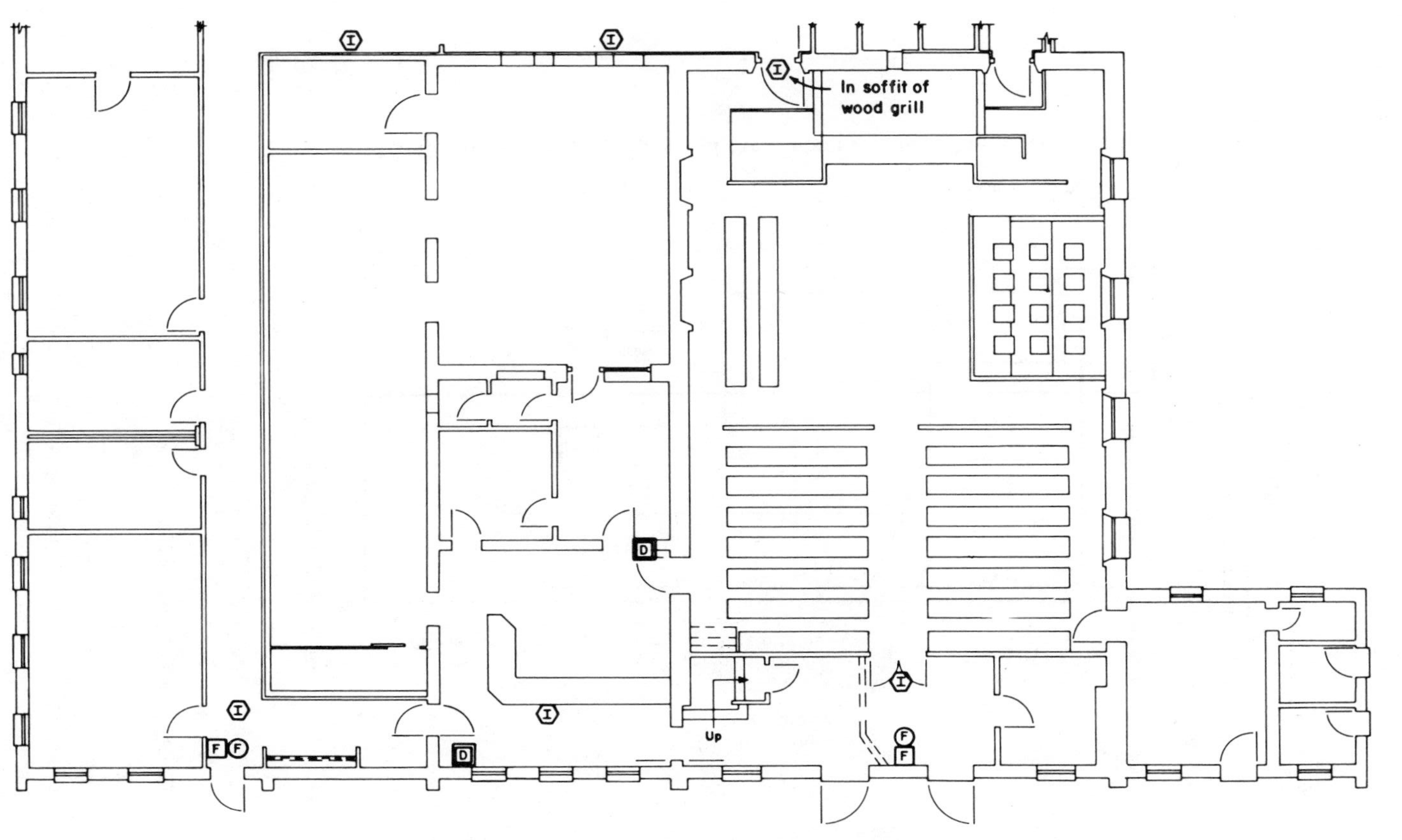

FIGURE 6-11. Floor plan of a small courthouse.

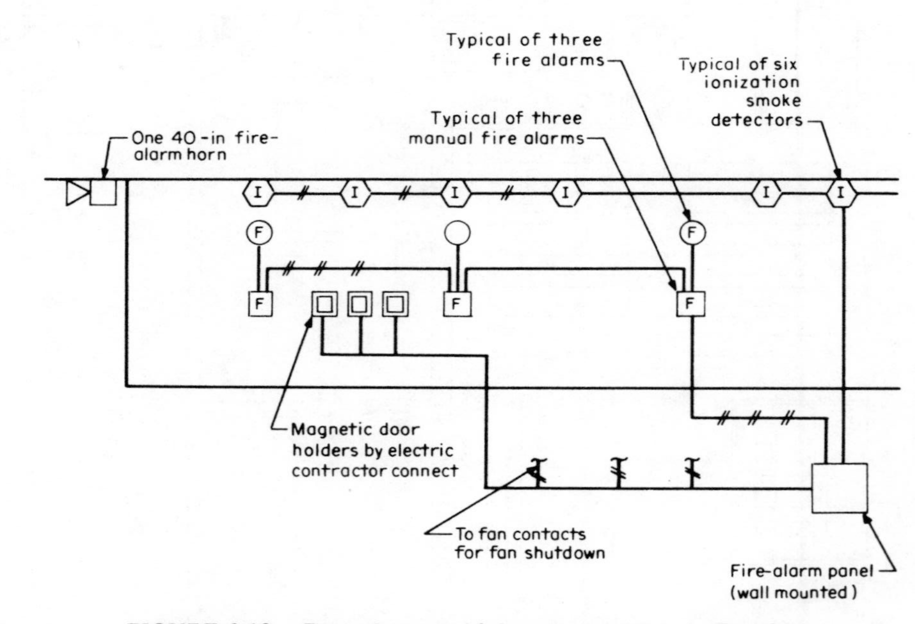

FIGURE 6-12. Riser diagram of the system shown in Fig. 6-11.

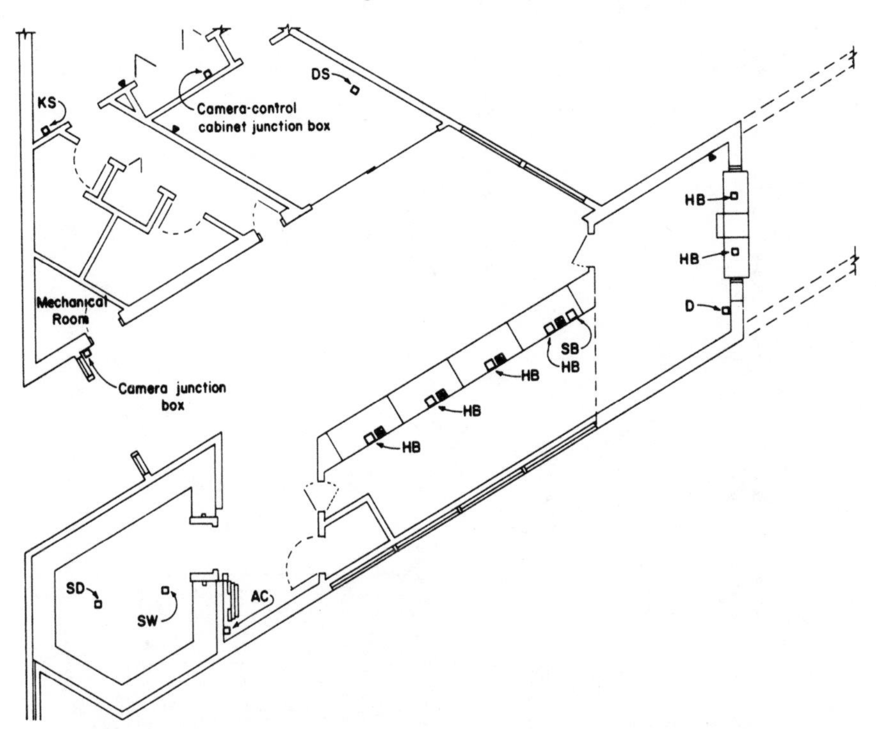

FIGURE 6-13. Floor plan of a branch bank showing the various outlets for the security system.

as discussed previously in this book. A riser diagram of this alarm system is shown in Fig. 6-12. Besides the normal components, note that this system utilizes magnetic contacts on each of the three air-handling units used for air-conditioning purposes. When any part of the fire-alarm system is activated, either manually or automatically, all fans are automatically shut down to stop the flow of air within the building. This action keeps the spread of fire to a minimum.

Figure 6-13 shows the floor plan of a branch bank along with the various outlets for the security system. Included are camera junction boxes, smoke detectors, sound receivers, alarm buttons, and so forth. They are shown on the floor plan only to indicate the approximate location of each. Details of wiring are not shown in this drawing, but the riser diagram in Fig. 6-14 indicates clearly how each outlet is to be installed.

INDUSTRIAL-ALARM SYSTEMS

Industrial security systems are very similar to residential and commercial systems in that they must detect, be controlled, and annunciate in the event of a break-in. However, they are more sophisticated and usually cover more territory.

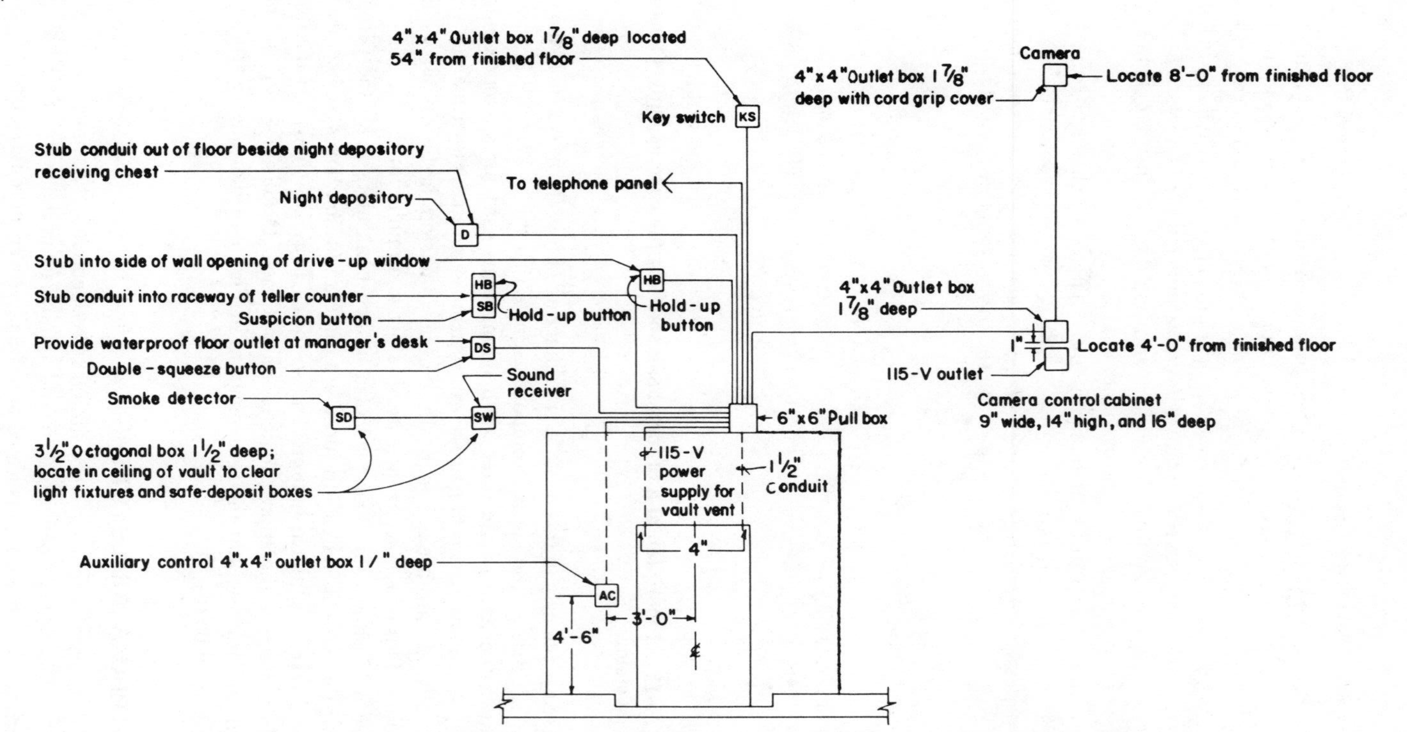

FIGURE 6-14. Riser diagram of the system shown in Fig. 6-13.

Fire-Alarm Systems

Also, fire-alarm systems are usually substantively different—requiring many unique controls to fit the various situations. For example, assume that a fire broke out in a paint-spraying area where flammable paint vapors were present. If the spray machines were allowed to continue operating after the fire had started, they would obviously feed the fire, making it almost impossible to extinguish. Therefore, some means of shutting off the paint sprayers must be incorporated into the fire-alarm system should the alarm sound.

Other machines, such as air-handling units, flammable chemical pumps, etc., must also be dealt with in the event of a fire. One way is to use relays in conjunction with the fire-alarm system so that if the alarm is activated, the relays will open the power-supply circuits which will shut down all machines and equipment.

Burglar-Alarm Systems

Most larger industrial plants utilize chain link fences to enclose the plant and surrounding grounds. Guards are then usually stationed at all entrances and exits. Their purpose is to prevent unauthorized personnel from entering the plant as well as to prevent pilferage from the plant's employees.

Where security is of the utmost importance, fence detectors are employed to detect any climbing or cutting of the fence. Watch dogs are sometimes used, as well as all the devices described previously in this book.

Closed-circuit television cameras are frequently used to monitor various areas of the plant and all are wired to a central location such as the guardroom where the areas can be checked at will by the guard on duty.

THE NUCLEAR ENVIRONMENT

The design of a security/fire-alarm in a nuclear reactor facility—designed for the generation of electrical power and employing enriched uranium or plutonium as a fuel element—is somewhat more complicated than designs for other facilities since many factors have to be considered; e.g., the type and quantity of nuclear phenomena that will be directed onto continuous thermistor sensors. Designers usually plan for an environment approximately 1.0×10^3 times more severe than that to which the sensor might normally be exposed.

Continuous thermistor sensors used for nuclear reactors are identical in construction to standard sensors with the exception that all-ceramic inserts are employed in the connectors.

The ceramic core of a typical continuous thermistor sensor can be thought of as an intrinsic semiconductor with an energy gap of approximately 1.1 eV. For all practical purposes, the sensor is an insulator at room temperature and normal radiation levels.

Conductivity in the sensor core increases by application of external energy to the system. Thus, considering all factors, the conductivity of the sensor will be an approximately exponential increasing function of applied external energy.

In analyzing the effect of radiation on the sensor, the most accurate approach is to relate the level of incident radiation to an equivalent temperature rise. This is done by relating the level of incident radiation to the number of MeV which can then be related to an increase in temperature.

The two types of radiation which could have a possible effect on the sensor are an integrated neutron dose which can be equated to an energy level of approximately 10.0 MeV and a total gamma flux equivalent to a deposited energy level of 500 MeV. Summation of these two levels gives an equivalent deposited energy level of 510 MeV. Since a temperature increase of 1°C is equivalent to an increase in deposited energy of 25.8 MeV, this level of radiation is equivalent to a temperature increase of 20°C.

It must be remembered that these are the radiation levels of the primary nuclear interface without benefit of shielding and as such are far in excess of the levels to which the sensor would normally be exposed.

HIGH-EXPANSION FOAM SYSTEM

High-expansion foam is an amalgamated aggregation of bubbles that are mechanically generated by the forced passage of air through a perforated stainless steel screen that has been wetted by an aqueous detergent solution. These foams provide a unique vehicle for the delivery of water to a fire and for the suppression of unignited vapors by a blanketing action.

The high-expansion foam produced for use on LNG (liquefied natural gas) fires and unignited spills will be generated by a 500:1 expansion of water and applied to an LNG spill trench at an application rate of 6.5 $ft^3/min/ft^2$ (34 $L/s/m^2$) of LNG surface area. These criteria result in a foam-application-rate parameter (flow rate divided by expansion ratio) of 0.013.

When applied to an LNG fire, high-expansion foam has the following effects:

1. It prevents air from reaching the fire when applied in sufficient quantity.

2. It reduces the oxygen content of the surrounding air by conversion of the water in the foam to steam.

3. It reduces the temperature of the fire by conversion of the water in the foam to steam.

When applied to an unignited LNG spill, a covering of high-expansion foam reduces the concentration of flammable vapors in the surrounding air by increasing the vapor boil-off temperature through insulation.

A typical application of the above system is used for an LNG trench area on Staten Island, NY. The trench is divided into three detection/protection zones which are designated zones A, B and C. A layout plan and details are shown in Figures 6-15 and 6-16.

ZONE A

Trench length: 550 ft (167 m)

LNG width at maximum spill depth: 19 ft (5.8 m)

Maximum LNG surface area: 10,450 ft^2 (971.9 m^2)

High-expansion foam generating capability required (based on 6.5 $ft^3/min/ft^2$ or 34 $L/s/m^2$ of LNG surface area): 67,925 ft^3/min (32 061 L/s)

Allowance for foam breakdown and weather: 25% (16,981 ft^3/min or 8 015 L/s)

Total generating capability required: 84,906 ft^3/min (40 076 L/s)

Based on these calculations Zone A will be protected by nine (9) 10,000 ft^3/min (4 719.3 L/s) generators.

Water Flow

Generator expansion ratio: 500:1

Water flow required at each generator: 20 ft^3/min (150 gal/min or 9.4 L/s)

Pressure required at each generator: 10 lb/in^2 (69 kPa)

Generator 1 to Distribution Tee

Required at generator: 10 lb/in^2 (69 kPa) at 150 gal/min (9.4 L/s)

Pipe size: 6 in (15 cm)

Flow item	*Length, ft (m)*
Pipe	200 (61)
Tee (side)	30 (9.2)

Total equivalent length: 230 ft (80.5 m)

Friction loss at maximum flow: 0.020 $lb/in^2/ft$ (457 Pa/m)

Pressure drop: 4.6 lb/in^2 (32 kPa)

Required at distribution tee: 14.6 lb/in^2 (101 kPa) at 1350 gal/min (85 L/s)

Distribution Tee to Regulator Valve Input

Required at distribution tee: 14.6 lb/in^2 (101 kPa) at 1350 gal/min (85 L/s)

Pipe size: 6 in (15 cm)

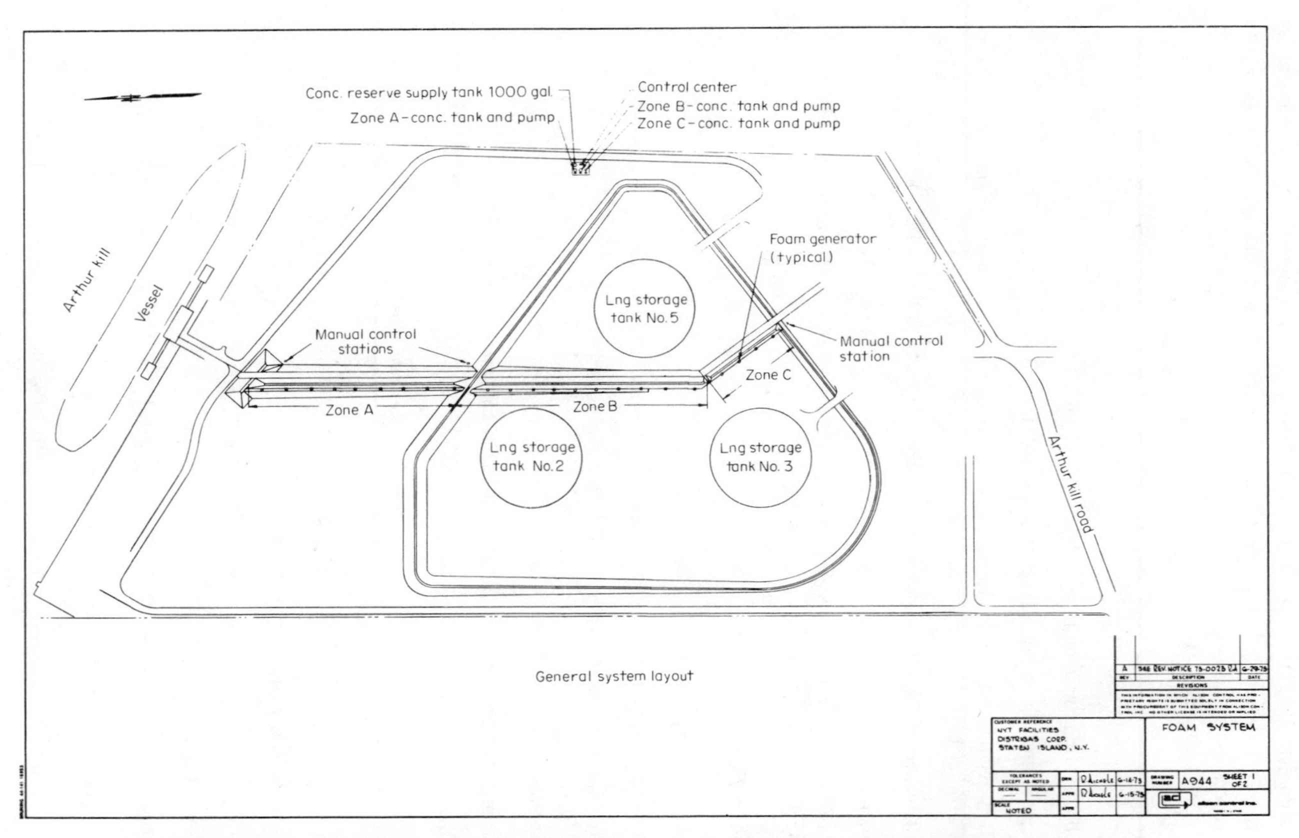

FIGURE 6-15. Site plan of an LNG layout that is protected by a high-expansion foam system.

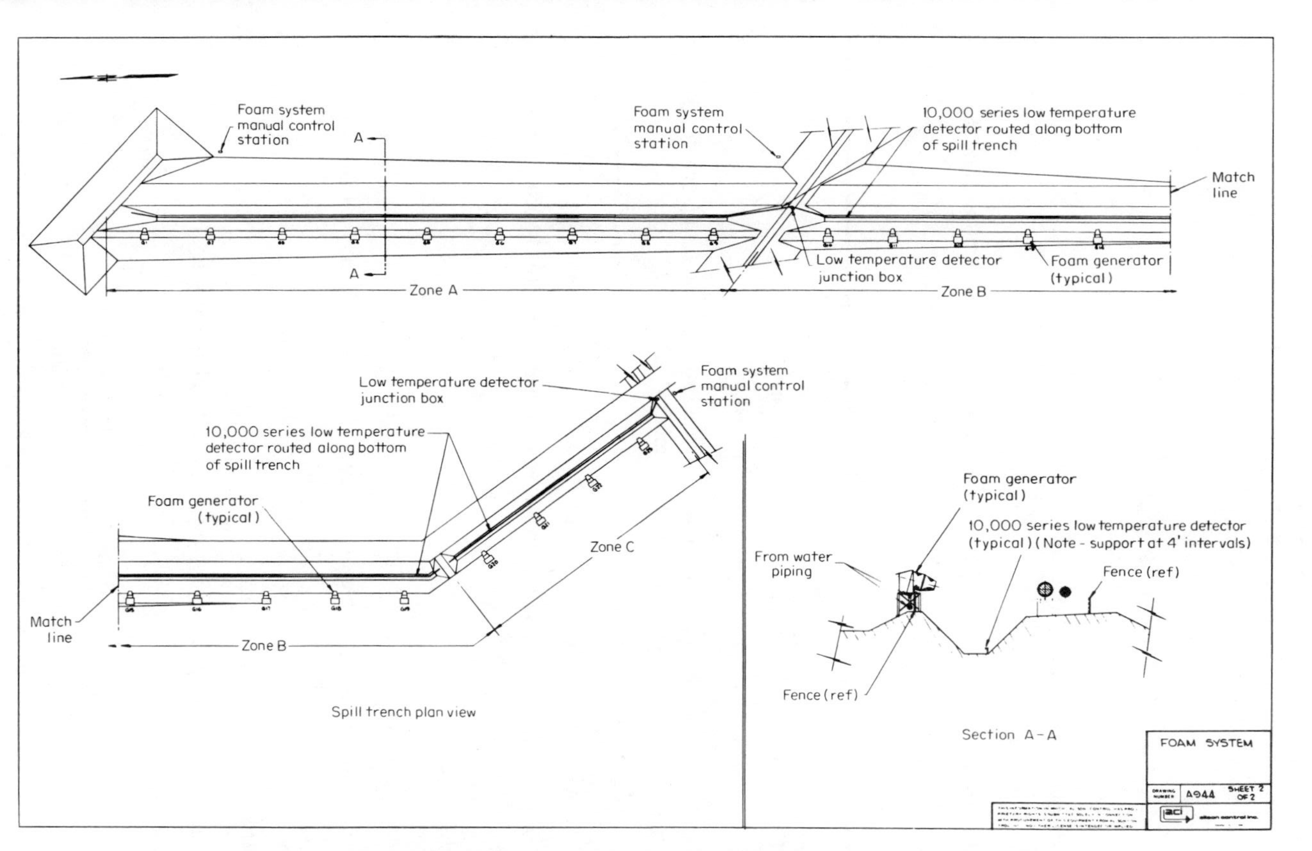

FIGURE 6-16. Sectional views and details of the system shown in Fig. 6-16.

Flow item	***Length, ft (m)***
Pipe	20 (6.1)

Total equivalent length: 20 ft (6.1 m)

Friction loss: 0.063 lb/in^2/ft (1.5 kPa/m)

Pressure drop: 1.26 lb/in^2 (8.7 kPa)

Elevation change: 3 ft (91 cm)

Elevation pressure drop: 1.3 lb/in^2 (8.9 kPa)

Valve differential pressure: 10 lb/in^2 (69 kPa)

Required at regulator valve input: 27.16 lb/in^2 (187.2 kPa) at 1350 gal/min (85 L/s)

ZONE B

Trench length: 600 ft (183 m)

LNG width at maximum spill depth: 19 ft (5.8 m)

Maximum LNG surface area: 11,400 ft^2 (1 060 m^2)

High-expansion foam generating capability required (based on 6.5 ft^3/min/ft^2 (34 L/s/m^2) of LNG surface area): 74,100 ft^3/min (34 975 L/s)

Allowance for foam breakdown and weather: 25% (18,525 ft^3/min or 8 744 L/s)

Total generating capability required: 92,625 ft^3/min (43,812 L/s)

Based on these calculations Zone B will be protected by ten (10) 10,000 ft^3/min (4 719.3 L/s) generators.

Water Flow

Generator expansion ratio: 500:1

Water flow required at each generator: 20 ft^3/min (150 gal/min or 9.4 L/s)

Pressure required at each generator: 10 lb/in^2 (69 kPa)

Generator 1 to Distribution Tee

Required at generator: 10 lb/in^2 (69 kPa) at 150 gal/min (9.4 L/s)

Pipe size: 6 in (15 cm)

Flow item	***Length, ft (m)***
Pipe	225 (68.6)
Tee (side)	30 (9.2)

Total equivalent length: 255 ft (77.8 m)

Friction loss at maximum flow: 0.021 lb/in^2/ft (475 Pa/m)

Pressure drop: 5.35 lb/in^2 (36.9 kPa)

Required at distribution tee: 15.35 lb/in^2 (105.8 kPa) at 1500 gal/min (94.5 L/s)

Distribution Tee to Regulator Valve Input

Required at distribution tee: 15.35 lb/in^2 (105.8 kPa) at 1500 gal/min (94.5 L/s)

Pipe size: 6 in (15 cm)

Flow item	Length, ft (m)
Pipe	20 (6.1)

Total equivalent length: 20 ft (6.1 m)

Friction loss: 0.075 $lb/in^2/ft$ (1.7 kPa/m)

Pressure drop: 1.50 lb/in^2 (10.3 kPa)

Elevation change: 3 ft (91 cm)

Elevation pressure drop: 1.3 lb/in^2 (8.9 kPa)

Valve differential pressure: 10 lb/in^2 (69 kPa)

Required at regulator valve input: 28.15 lb/in^2 (194.1 kPa) at 1500 gal/min (94.5 L/s)

ZONE C

Trench length: 230 ft (70.2 m)

LNG width at maximum spill depth: 19 ft (5.8 m)

Maximum LNG surface area: 4370 ft^2 (406.4 m^2)

High-expansion foam generating capability required (based on 6.5 $ft^3/min/ft^2$ or 34 $L/s/m^2$ of LNG surface area): 28,405 ft^3/min (13 407 L/s)

Allowance for foam breakdown and weather: 25% (7101 ft^3/min or 3 352 L/s)

Total generating capability required: 35,506 ft^3/min (16 759 L/s)

Based on these calculations Zone C will be protected by four (4) 10,000 ft^3/min (4 719.3 L/s) generators.

Water Flow

Generator expansion rate: 500:1

Water flow required at each generator: 20 ft^3/min (150 gal/min or 9.45 L/s)

Pressure required at each generator: 10 lb/in^2 (69 kPa)

Generator 1 to Distribution Tee

Required at generator: 10 lb/in^2 (69 kPa) at 150 gal/min (9.45 L/s)

Pipe size: 4 in

Flow item	***Length, ft (m)***
Pipe	100 (30.5)
Tee (side)	20 (6.1)

Total equivalent length: 120 ft (36.5 m)

Friction loss at maximum flow: 0.028 lb/in^2/ft (633 Pa/m)

Pressure drop: 3.36 lb/in^2 (23.2 kPa)

Required at distribution tee: 13.36 lb/in^2 (92.12 kPa) at 600 gal/min (37.8 L/s)

Distribution Tee to Regulator Valve Input

Required at distribution tee: 13.36 lb/in^2 (92.12 kPa) at 600 gal/min (37.8 L/s)

Pipe size: 4 in (10 cm)

Flow item	***Length, ft (m)***
Pipe	20 (6.1)

Total equivalent length: 20 ft (6.1 m)

Friction loss: 0.101 lb/in^2/ft (2.28 kPa/m)

Pressure drop: 2 lb/in^2 (13.8 kPa)

Elevation change: 3 ft (91 cm)

Elevation pressure drop: 1.3 lb/in^2 (8.9 kPa)

Valve differential pressure: 10 lb/in^2 (69 kPa)

Required at regulator valve input: 26.66 lb/in^2 (183.8 kPa) at 600 gal/min (37.8 L/s)

The water required for all three zones must equal 28.15 lb/in^2 (194.1 kPa) at 3450 gal/min (217.4 L/s). Water is available to the above-ground main at both ends.

The 10-in (25-cm) line supplying the high-expansion foam system is aboveground and normally must be kept dry. To isolate it from the water in the fire loop, an electrically operated solenoid valve is employed at each end so that when any zone is actuated both these valves will be opened, admitting water from both ends of the pipe.

The following calculations indicate the required pressure and flow from one end of the above-ground main with the other connection inoperative. The calculations present the pressure/flow calculation to the OS and Y valve which isolates the high-expansion foam system.

Required: 28.15 lb/in^2 (194.1 kPa) at 3450 gal/min (217.4 L/s)

Pipe size: 10 in (25 cm)

Flow item	*Length, ft (m)*
Pipe	690 (210.5)
Check valve (swing, equivalent length)	14 (4.3)
Solenoid valve (equivalent length)	161 (48.9)

Strainer (10 in or 25 cm): 3.5 lb/in^2 actual drop

Total equivalent length: 865 ft (264 m)

Friction loss: 0.028 lb/in^2/ft (633 Pa/m)

Pressure drop: 24.22 lb/in^2 (166.9 kPa)

Required at foam system OS and Y valve: 53 lb/in^2 (365.4 kPa) at 3450 gal/min (217.4 L/s)

Concentrate for each of the three zones is stored in three fiberglass tanks which are heated by trace heaters controlled by the same control-center thermostat that controls generator heating. Each tank contains sufficient concentrate for two 10-minute foam producing periods. A fourth tank containing reserve concentrate may be used to refill any of the three other tanks by means of a solenoid valve matrix and a fill pump.

The concentrate from each zone's tank is moved through a network of stainless steel pipes by means of a redundant pumping network. If the primary pump should fail to operate, a flow switch senses the lack of flow and the secondary pump is automatically started.

Concentrate pressure is controlled by a bypass-type pressure regulator and concentrate flow is controlled by a metering orifice. The pressure regulator will be set to overcome the water pressure at the injection assembly.

ZONE A

Generator capacity: 90,000 ft^3/min (42 480 L/s)

Water requirement: 1350 gal/min (85.1 L/s)

Injection ratio: 3%

Concentrate flow: 40.5 gal/min (2.55 L/s)

Concentrate required: 405 gal (1 533 L) per discharge

Total concentrate tank capacity: 810 gal (3 066 L)

Required at injection point: 40.5 gal/min (2.55 L/s) at 17.16 lb/in^2 (118 kPa)

ZONE B

Generator capacity: 100,000 ft^3/min (47 200 L/s)

Water requirement: 1500 gal/min (94.5 L/s)

Injection ratio: 3%

Concentrate flow: 45 gal/min (2.8 L/s)

Concentrate required: 450 gal (1 703 L) per discharge

Total concentrate tank capacity: 900 gal (3 407 L)

Required at injection point: 45 gal/min (2.8 L/s) at 18.15 lb/in^2 (125 kPa)

ZONE C

Generator capacity: 40,000 ft^3/min (18 880 L/s)

Water requirement: 600 gal/min (37.8 L/s)

Injection ratio: 3%

Concentrate flow: 18 gal/min (1.13 L/s)

Concentrate required: 180 gal (681 L) per discharge

Total concentrate tank capacity: 360 gal (1 363 L)

Required at injection point: 18 gal/min (1.13 L/s) at 16.66 lb/in^2 (115 kPa)

The system described above was designed and furnished by Alison Control, Inc., 35 Daniel Road West, Fairfield, NJ 07006. For additional information, you may write to them directly.

The basic actuation means employed with the Alison Control high-expansion foam system are combustible vapor detectors and explosion-proof manual-control stations. Normally open output contacts of the combustible vapor detectors are run in a four-wire three-loop configuration back to the foam-control center where a closure of any output contact will discharge foam into the zone indicating an alarm.

Combustible-vapor detector alarms are annunciated both in the field at the foam-control center and in the control room. The extension wiring loops of the combustible-vapor alarm contacts are supervised and annunciated for continuity. The Alison Control manual-alarm stations are run in a four-wire three-loop configuration back to the foam-control center where a momentary closure of any manual start switch will discharge foam into the proper zone. FOAM-ON conditions are annunciated both at the control center and in the control room. The extension wiring loops of the manual control stations are supervised and annunciated for continuity.

The entire high-expansion foam system is operated from and supervised by a single foam-control center located outside of the hazardous area. The operation of the electronic controls can be best understood by examining the three operation modes it may assume. These are the standby mode, the alarm response mode, and the optional foam monitoring mode.

Standby Mode

The standby mode is defined as all system components in their normal condition with all alarms in their normal condition and operating power available to the system. It is in this mode that the system will spend the majority of its time.

The high-expansion foam system designed by Alison Control employs a redundant-signal-loop, pulse-coded, modulation-control method to electrically actuate all system components external to the foam-control center. These components include all generator motors, all pump motors, all water-control valves, and all solenoid-control valves.

A pulse-coded modulation (PCM) power-control system offers significant advantages in installation costs when any electrical items must be actuated from a single remote-control location.

In a PCM system a single power-distribution loop is run connecting all items that operate from the same voltage in parallel but isolated from each other by a switching network. This single power loop drastically reduces the number of power feeds required. For example, the 23 generators required for this system would normally have to be supplied by 69 wires for operation from a three-phase source. With a PCM system only three power wires are required. These wires are connected in a single continuous loop to each motor in turn.

Each remote item in a PCM system is assigned a binary code which uniquely identifies it. The switching device which isolates any remote item from the power loop is actuated only by the presence of this code. In operation, all identifying codes along with a turn-on or turn-off code are transmitted serially along a supervised two-wire control loop.

A typical code sequence involves transmission of a DATA INITIATE bit, a TURN ON command followed by transmission of the codes of all items that are to turn on, and finally a DATA COMPLETE bit. This sequence will turn on all desired items. A similar sequence containing a TURN OFF command is used for system shutdown.

Normally only a single signal loop is employed in any PCM system for transmission of RUN/STOP commands. However, due to the nature of this system, a redundant-signal-loop system is employed.

The redundant-signal-loop system utilizes two completely separate signal loops employing complementary coding. Signals are transmitted on both loops simultaneously and each signal is completely capable of system control.

The high-expansion foam system has four loops: two power loops and two signal loops. One power loop is a 3-wire, three-phase 230/460-V loop for operation of all motors. The other power loop is a 120-Vac signal-phase loop for operation of all solenoids.

All loops in the pulse-coded modulation system are supervised for continuity and any trouble conditions are annunciated at the control center and at the control-room console.

Weather protection One of the most important considerations, when the system is operating in its standby mode, is the weather protection of all exposed equipment including, most importantly, protection of the generator screen and air intake from icing conditions.

Each generator is equipped with 500 W of heating power which is derived directly from the operate power loop. Three hundred watts of this power is used to protect the foam-forming screen from ice and the remaining 200 W deices the air-intake screen and venturi.

The heating is controlled by a thermostat located at the control center which will transmit a heater ON/OFF code along the signal loop.

A separate heating loop is provided for the concentrate tank and all filled piping. Heating of all filled water pipes, to a depth sufficient to overcome the danger of frost, is provided. All other electrical controls are enclosed in weatherproof enclosures sufficient for the environment for which they are installed.

Alarm-Response Mode

Any zone will go into its alarm response (operate) mode when it receives either a contact closure from a combustible vapor detector or a manual start signal.

When this happens, the zone in alarm will produce foam until either a STOP button located either in the remote-control station or on the foam-control station or in the control room is operated or until the first foam-producing period as sensed by a level switch in the concentrate tank is completed.

If the system stops as a result of a STOP command generated by a concentrate level switch, a second equal foam-producing period may be initiated by operation of a level override switch and a new START command.

Monitoring Mode (Optional)

The high-expansion foam system can be equipped with a monitoring mode which, upon receipt of an alarm signal from either a combustible vapor detector or a low-temperature detector, will cause a minimum 3-ft (0.9-m)-high blanket of foam to be deposited on the LNG to reduce the vapor concentrations. At the end of a 1-minute foam producing period, the foam system will stop and wait for a delay period that is adjustable between 1 and 10 minutes.

At the end of the chosen delay period, the foam system is reactuated automatically for a period of 1 minute to replenish the foam blanket. An operator may override a monitoring-mode operation at any time by operation of a manual START button.

Concentrate refill Since operation in the monitoring mode may extend over a long period of time, it is possible to refill any of the concentrate tanks at any time from either the control room or the control center.

This is done by selecting the tank to be filled by means of a selector switch and by operating a refill-pump run switch which will pump concentrate from the reserve tank to the selected tank.

Since the levels of all concentrate tanks are annunciated both on the control center and in the control room, pumping can cease when the tank selected has been completely refilled.

7

READING ELECTRICAL BLUEPRINTS

Those persons involved in the design and installation of security/fire-alarm systems must know how to interpret drawings, wiring diagrams, and other supplementary information found in drawings and written specifications. This chapter is designed to provide the reader with a review of electrical blueprint reading as it relates to the security/fire-alarm field.

TYPES OF ELECTRICAL DRAWINGS

In order to be able to "read" any of the drawings found in the design and installation of security/fire-alarm systems, one must first become familiar with the meaning of the various symbols, lines, and abbreviations used on the drawings and learn how to interpret the message conveyed by each one.

The types of electrical drawings usually fall into one of the following categories:

1. Electrical construction drawings.
2. Single-line block diagrams.
3. Schematic wiring diagrams.

Electrical construction drawings show the physical arrangement and views of specific electrical equipment. These drawings give all the plan views, elevation views, and other details necessary to construct the item. For example, Fig. 7-1 shows a pictorial sketch of a wire trough (auxiliary gutter). One side of the trough is labeled "top", one labeled "front," and another labeled "end."

This same trough is represented in another form in Fig. 7-2. The drawing labeled "top" is what one sees when one views the trough directly from above; the

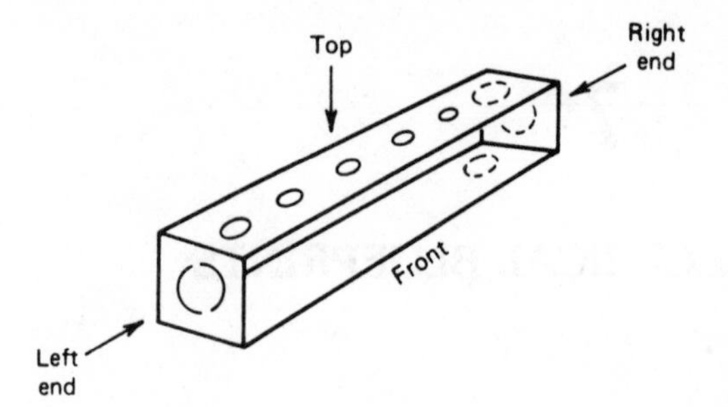

FIGURE 7-1. Pictorial drawing of a wire trough.

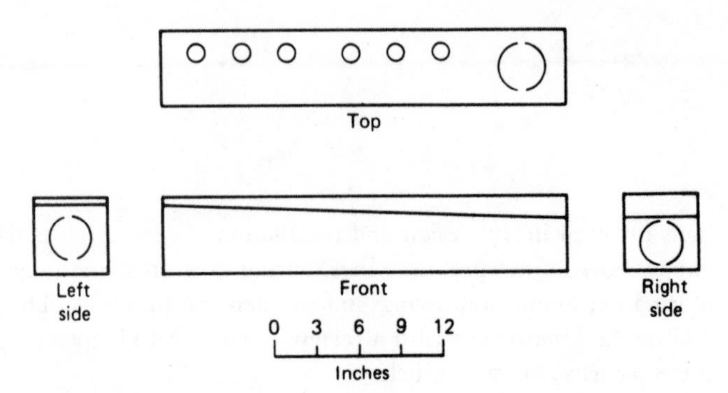

FIGURE 7-2. Top, front, and side views of the wire trough shown in Fig. 7-1.

ones labeled "left side" or "right side" are the views from the sides; and the drawing labeled "front" shows what the gutter looks like when one views the panel directly from the front of it.

The width of the trough is shown by the horizontal lines on the top view and the horizontal lines of the front view. The height is shown by the vertical lines of both the front and the side views; while the depth is shown by the vertical lines of the top view and the horizontal lines of the side views.

The four drawings in Fig. 7-2 clearly give the shape of the wire trough, but the drawings alone would not enable a worker to construct the trough because there is no indication of its size. There are two common methods to indicate the actual length, width, and height of the wire trough. One is to draw all of the views to some given scale, such as 1½ in = 1 ft-0 in. This means that 1½ in (3.75 cm) on the drawing represents 1 ft (30 cm) in the actual construction of the wire trough. The scale is then indicated as shown at the bottom of Fig. 7-2. The second method is to give dimensions on the drawing such as those shown in Fig. 7-3.

Drawings like the ones just covered are used mainly by equipment manufacturers. Designers and installers of security/fire-alarm systems will more often run

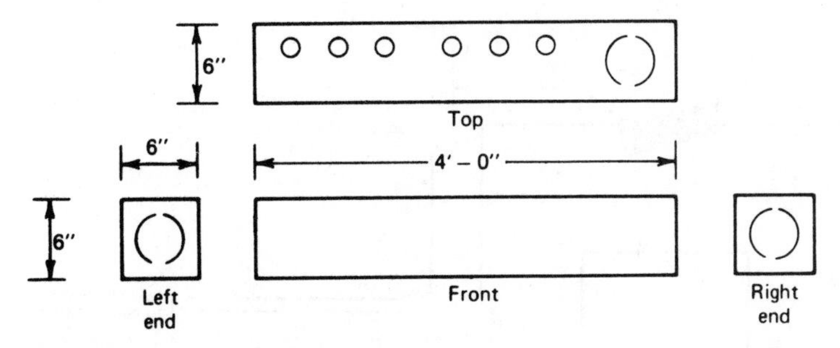

FIGURE 7-3. The views shown Fig. 7-2; the drafter has used an alternative method of showing dimensions.

across construction drawings like the one shown in Fig. 7-4. This type of construction drawing is normally used to supplement building drawings for a special system installation and is often referred to as a *detail drawing.*

Electrical diagrams intend to show, in diagrammatic form, electrical components and their related connections. In diagrams, electrical/electronics symbols are used extensively to present the various components. Lines are often used to connect these symbols, indicating the size, type, and number of wires that are necessary to complete the electrical circuit.

Security/fire-alarm system installers will often come into contact with *single-*

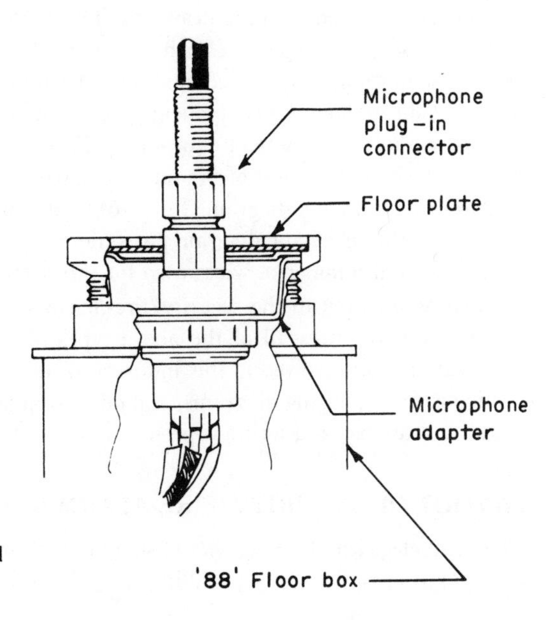

FIGURE 7-4. Typical detail drawing.

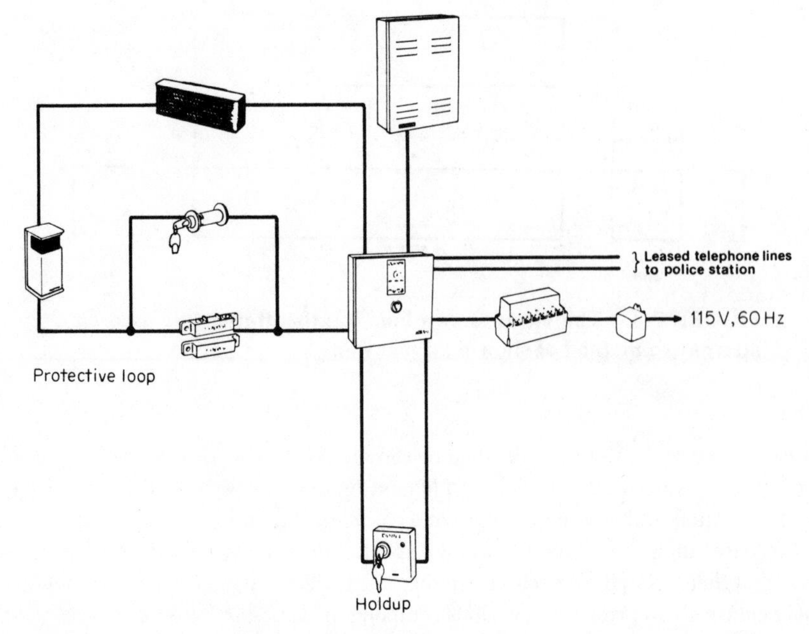

FIGURE 7-5. Single-line block diagram of a commercial intrusion system.

line block diagrams. These are used extensively to indicate the arrangement of systems on shop and working drawings. The *commercial intrusion system* in Fig. 7-5, for example, is typical of such drawings. This particular drawing shows all of the system's major equipment and related components, as well as the connecting lines to indicate how the components are connected to the system. Numerals are used to identify each piece of equipment. This drawing shows the components in pictorial fashion, but most of the time, they will appear in symbol form.

A schematic wiring diagram (Fig. 7-6) is similar to a single-line block diagram except that the schematic diagram gives more detailed information and shows the actual size and number of wires used for the electrical connections.

Anyone involved in the security/fire-alarm system industry, in any capacity, frequently encounters all of the above types of drawings. Therefore, it is very important for all involved in this industry to fully understand electrical and electronic drawings, wiring diagrams, and other supplementary information found in working drawings and written specifications.

LAYOUT OF SECURITY/FIRE-ALARM SYSTEM DRAWINGS

The ideal electrical drawing should show in a clear, concise manner exactly what is required of the workers installing or maintaining the system. The amount of

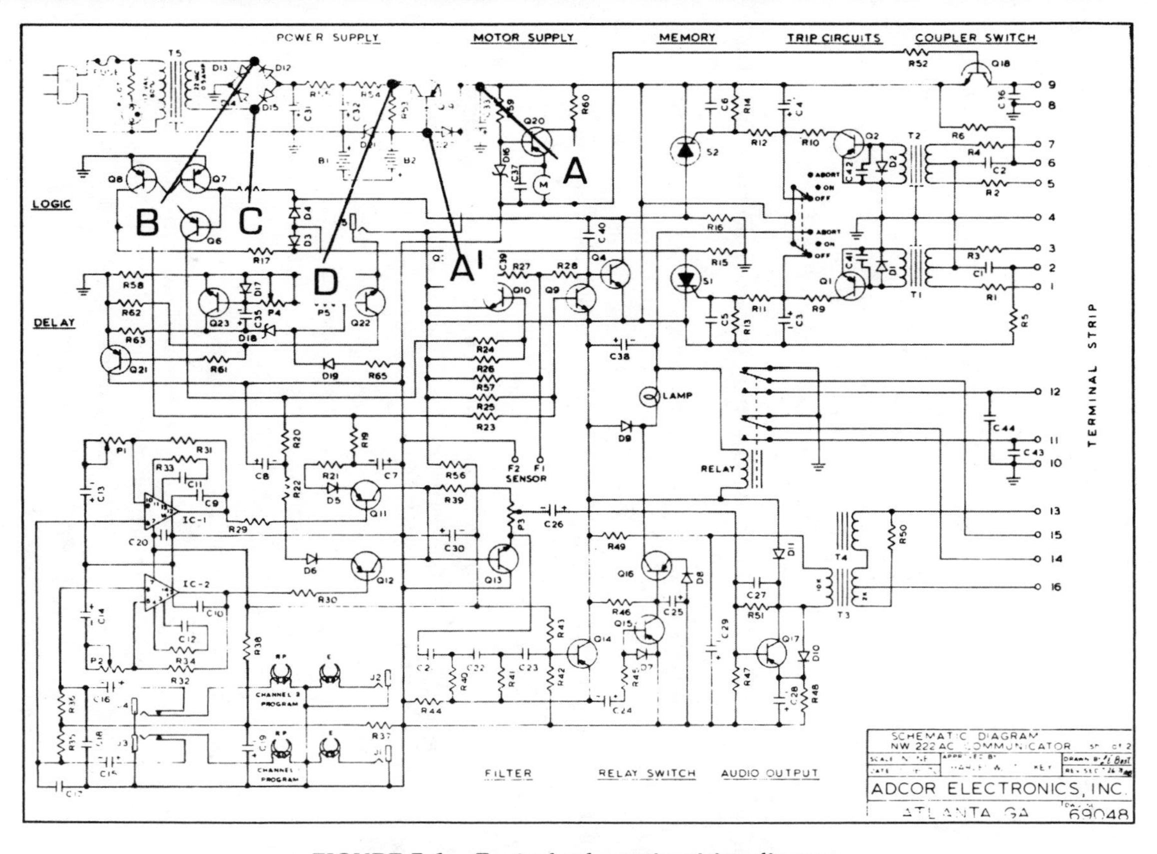

FIGURE 7-6. Typical schematic wiring diagram.

data shown on such a drawing should be sufficient, but not overdone. Unfortunately, this is not always the case. The quality of electrical drawings vary from excellent, complete, and practical, to just the opposite. In some cases, the design may be so incomplete that the installers will have to supplement it—if not completely design it—prior to estimating the cost of the project or beginning installation of the system.

In general, a good working drawing of a security/fire-alarm system should contain floor plans of each floor of the building to show the physical arrangement of the various components in the building, riser diagrams or schematic diagrams to show how the equipment is connected, schedules to indicate the size and type of equipment, and large-scale detailed drawings for special or unusual portions of the installation. A legend of electrical/electronic symbols should be included on the drawing to explain the meaning of every symbol and line used on the drawing. Anything that cannot be shown by symbols and lines should be clarified with neatly lettered notes or explained in the written specifications. The scale to which the drawings are prepared is also important; they should be as large as practical and where dimensions are to be held to extreme accuracy, dimension lines should be added.

The following steps are usually necessary in preparing a good set of working drawings and specifications:

1. The engineer or designer meets with the architect and owner to discuss the security needs of the building in question and also to discuss various recommendations made by all parties.

2. Once the data in no. 1 above is agreed upon, an outline of the architect's floor plan is drawn on tracing paper and several prints of this floor-plan outline are made.

3. The security/fire-alarm system is then laid out on the drawings as discussed in earlier chapters. All detectors, sounding devices, controls, etc., are located on the drawing, and each is identified by a symbol or note.

4. Schedules are used to identify various pieces of equipment.

5. Wiring diagrams are made to show the workers how various electrical components are to be connected. An electrical/electronic symbol list should also be included to identify the symbols used on the drawings.

6. Various large-scale details are included, if necessary, to show exactly what is required of the workers.

7. Written specifications are then made to give a description of the materials and the installation methods.

If these steps are properly taken in preparing a set of working drawings for security/fire-alarm systems, they will be detailed and accurate enough for a more rapid and accurate cost estimate, as well as a first-class installation.

ELECTRICAL/ELECTRONIC GRAPHIC SYMBOLS

Since electrical drawings must be prepared by electrical drafters in a given time period to stay within the allotted budget, symbols are used to simplify the work. In turn, a knowledge of electrical/electronic symbols must also be acquired by anyone who must interpret and work with the drawings.

Most engineers, designers, and drafters use symbols adopted by the American National Standards Institute (ANSI, formerly USASI) for use on electrical and electronics drawings. However, many designers and drafters frequently modify these symbols to suit their own particular requirements for the type of work they normally encounter. For this reason, most drawings have a symbol list or legend drawn and lettered on each set of working drawings.

The symbol list in Fig. 7-7 represents a good list for use on electrical drawings, while the one in Fig. 7-8 should suffice for most electronics drawings. Those who work with security/fire-alarm systems will encounter the symbols in both of these lists.

It is evident from these lists that many of the symbols have the same basic form, but their meanings are different because of some slight difference in the symbol. For example, the outlet symbols in Fig. 7-9 have the same basic form—a circle. However, the addition of a line or a dot to the circle, or perhaps a letter or note, gives each an individual meaning. Therefore, a good procedure to follow in learning the different symbols is first to learn the basic form and then to apply the variations for obtaining different meanings.

Some of the symbols listed utilize abbreviations to obtain their meaning, such as WP for weatherproof and E for emitter, C for collector, etc. Others are simplified pictographs, such as for a double floodlight fixture or for loudspeaker or for general antenna.

In some cases, the symbols are combinations of abbreviation and pictographs, such as F for fusible safety switch, and for PNP transistor.

SCHEDULES

A schedule, as related to drawings of security/fire-alarm systems, is a systematic method of presenting notes or lists of equipment on a drawing in tabular form. When properly organized and thoroughly understood, schedules not only are powerful timesaving methods for drafters, but also save security-system contractors and their workers much valuable time in preparing the estimate and installing the equipment in the field.

ELECTRICAL SYMBOLS

SWITCH OUTLETS

Single-Pole Switch	S
Double-Pole Switch	S_2
Three-Way Switch	S_3
Four-Way Switch	S_4
Key-Operated Switch	S_K
Switch and Fusestat Holder	S_FH
Switch and Pilot Lamp	S_P
Fan Switch	S_F
Switch for Low-Voltage Switching System	S_L
Master Switch for Low-Voltage Switching System	S_{LM}
Switch and Single Receptacle	⊖S
Switch and Duplex Receptacle	⊖S
Door Switch	S_D
Time Switch	S_T
Momentary Contact Switch	S_{MC}
Ceiling Pull Switch	Ⓢ
"Hand-Off-Auto" Control Switch	HOA
Multi-Speed Control Switch	M
Push Button	•

RECEPTACLE OUTLETS

Where weather proof, explosion proof, or other specific types of devices are to be required, use the upper-case subscript letters. For example, weather proof single or duplex receptacles would have the uppercase WP subscript letters noted alongside of the symbol. All outlets should be grounded.

Single Receptacle Outlet	⊖
Duplex Receptacle Outlet	⊖
Triplex Receptacle Outlet	
Quadruplex Receptacle Outlet	
Duplex Receptacle Outlet-Split Wired	
Triplex Receptacle Outlet-Split Wired	
250 Volt Receptable Single Phase Use Subscript Letter to Indicate Function (DW-Dishwasher; RA-Range, CD - Clothes Dryer) or numeral (with explanation in symbol schedule)	
250 Volt Receptacle Three Phase	
Clock Receptacle	Ⓒ
Fan Receptacle	Ⓕ
Floor Single Receptacle Outlet	
Floor Duplex Receptacle Outlet	
Floor Special-Purpose Outlet	*
Floor Telephone Outlet - Public	
Floor Telephone Outlet - Private	

Example of the use of several floor outlet symbols to identify a 2, 3, or more gang floor outlet:

Underfloor Duct and Junction Box for Triple, Double or Single Duct System as indicated by the number of parallel lines.

Example of use of various symbols to identify location of different types of outlets or connections for underfloor duct or cellular floor systems:

Cellular Floor Header Duct

*Use numeral keyed to explanation in drawing list of symbols to indicate usage.

FIGURE 7-7. List of electrical symbols.

CIRCUITING

Wiring Exposed (not in conduit) — E —

Wiring Concealed In Ceiling or Wall

Wiring Concealed in Floor

Wiring Existing*

Wiring Turned Up

Wiring Turned Down

Branch Circuit Home Run to Panel Board. (2 1)

Number of arrows indicates number of circuits. (A number at each arrow may be used to identify circuit number.)**

BUS DUCTS AND WIREWAYS

Trolley Duct*** (T T)

Busway (Service, Feeder, or (Plug-in)*** (B B)

Cable Trough Ladder or Channel*** (C C)

Wireway*** (W W)

PANELBOARDS, SWITCHBOARDS AND RELATED EQUIPMENT

Flush Mounted Panelboard and Cabinet***

Surface Mounted Panelboard and Cabinet***

Switchboard, Power Control Center, Unit Substations (Should be drawn to scale)***

Flush Mounted Terminal Cabinet (In small scale drawings the TC may be indicated alongside the symbol)*** (TC)

Surface Mounted Terminal Cabinet (In small scale drawings the TC may be indicated alongside the symbol)*** (TC)

Pull Box (Identify in relation to Wiring System Section and Size)

Motor or Other Power Controller (May be a starter or contactor)***

Externally Operated Disconnection Switch***

Combination Controller and Disconnection Means***

POWER EQUIPMENT

Electric Motor (HP as indicated) (1/4)

Power Transformer

Pothead (Cable Termination)

Circuit Element, e.g., Circuit Breaker (CB)

Circuit Breaker

Fusible Element

Single-Throw Knife Switch

Double-Throw Knife Switch

Ground

Battery

Contactor (C)

Photoelectric Cell (PE)

Voltage Cycles, Phase Ex: 480/60/3

Relay (R)

Equipment (Connection (as noted)

---

*Note. Use heavy-weight line to identify service and feeders. Indicate empty conduit by notation CO (conduit only).

**Note: Any circuit without further identification indicates two-wire circuit. For a greater number of wires, indicate with cross lines, e.g.

3 wires; 4 wires, etc.

Neutral wire may be shown longer. Unless indicated otherwise, the wire size of the circuit is the minimum size required by the specification. Identify different functions, of wiring system. e.g., signalling system by notation or other means.

***Identify by Notation or Schedule

FIGURE 7-7. *(Cont.)*

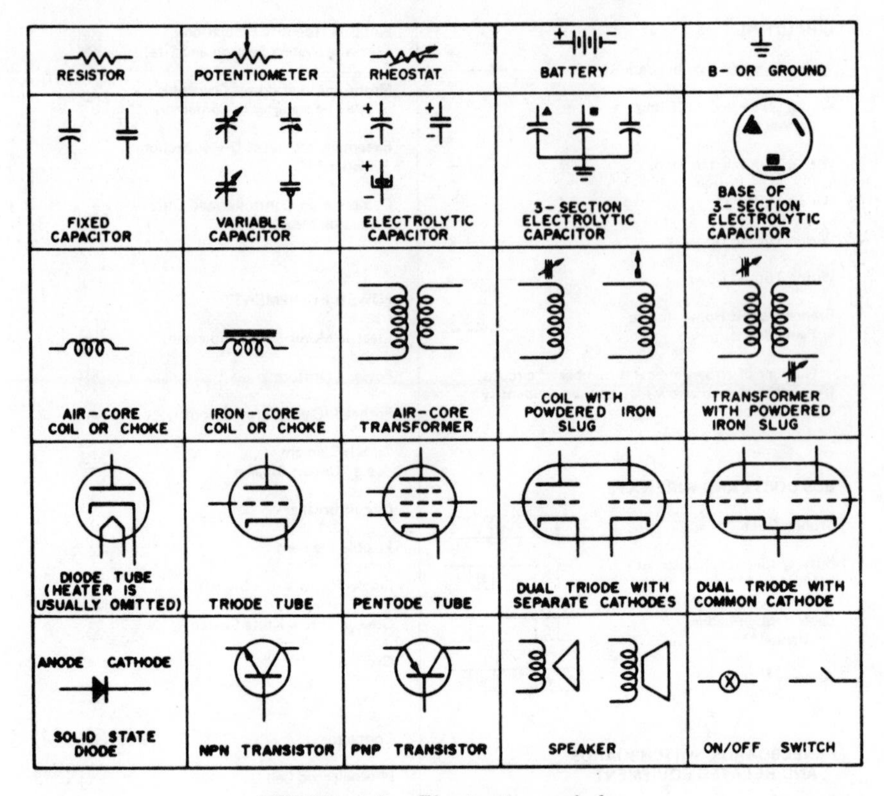

FIGURE 7-8. Electronic symbols.

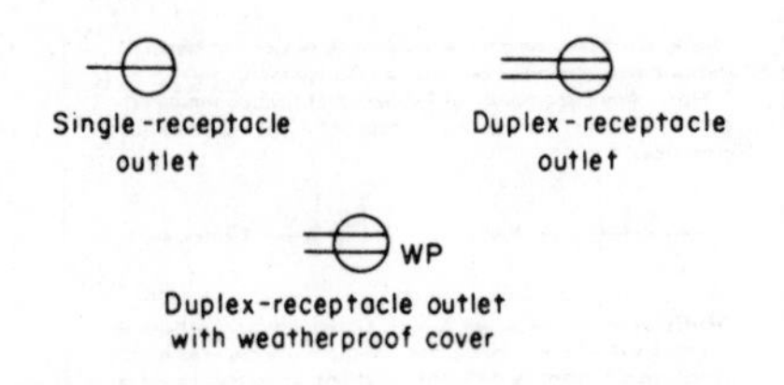

FIGURE 7-9. Various types of outlet symbols.

Capacitors	*Resistors (10%, 0.5W)*
C1 0.1μf 100V Orange Drop	R1 1 MΩ
C2 0.1μf 100V Orange Drop	R2 1 MΩ
C3 30μf 25V Electrolytic	R3 1 MΩ
C5 33μf 25V Electrolytic	R4 1 MΩ
C6 33μf 25V Electrolytic	R5 100 kΩ
C7 0.01μf 500V Disc Ceramic	R6 10 kΩ
C8 33μf 25V Electrolytic	R7 10 kΩ
C9 0.1μf 100V Orange Drop	R8 10 kΩ
C10 0.1μf 100V Orange Drop	R9 3.3 Ω
C11 0.1μf 100V Orange Drop	R10 3.3 Ω
Diodes	*Transistors*
D1 1N4149	Q1 MPS3704
D2 1N4149	Q2 2N4126
D3 1N4149	Q3 MPS3704
D4 1N4149	
D5 1N964B	
D6 1N4149	
D7 1N4149	
D8 1N4149	
Relays	*Power supply*
S1 SIGMA 65FPIA 6Vdc	PS 9Vdc at 200 mA—O.C.V. 12V max.
S2 INRESCO APC Form A, 1500T #34	
	Battery
	SPS EVEREADY

FIGURE 7-10. Schedule listing parts for a security/fire-alarm system.

Sometimes schedules are omitted on the drawings and the information is contained in the written specifications instead. It is very time-consuming to comb through page after page of written specifications, and workers do not always have access to the specifications while working; however, they usually do have access to the working drawings at all times.

The schedules in Figs. 7-10 and 7-11 are typical of those used on drawings pertaining to the installation of security/fire-alarm systems.

Channel 1	*Channel 2*	*Check-point*	*Proper voltage reading, dc*	*Notes*
Tripped	Not tripped	Q R	−7 to −9V −7 to −9V	1. Cross talk while channel one (two) is running may be caused by a "leaky" Q12 (Q11). The presence of a voltage drop across R30 (R29) also indicates a "leaky" Q12.
		S T U	−17 to −18V −12 to −13V −7 to −9V	2. If point "G" ("H") is not at zero volts when channel one (two) is tripped, Q5 and/or Q7 (Q6 and/or Q8) may be defective.
		V W	−22V 0	3. If point "H" ("G") is lower than approximately −22 Vdc while channel one (two) is running, Q6 and/or Q8 (Q5 and/or Q7) may be defective.
		X I J	−22V −20V 0	4. Refer to page 14B for repair instructions.
Not tripped	Tripped	Q	−7 to −9V	
		R S T U	−7 to −9V −17 to −18V −12 to −13V −22V	
		V	−7 to −9V	
		W	−22V	
		X	0	
		I	0	
		J	−20V	

FIGURE 7-11. Schedule used for troubleshooting procedures.

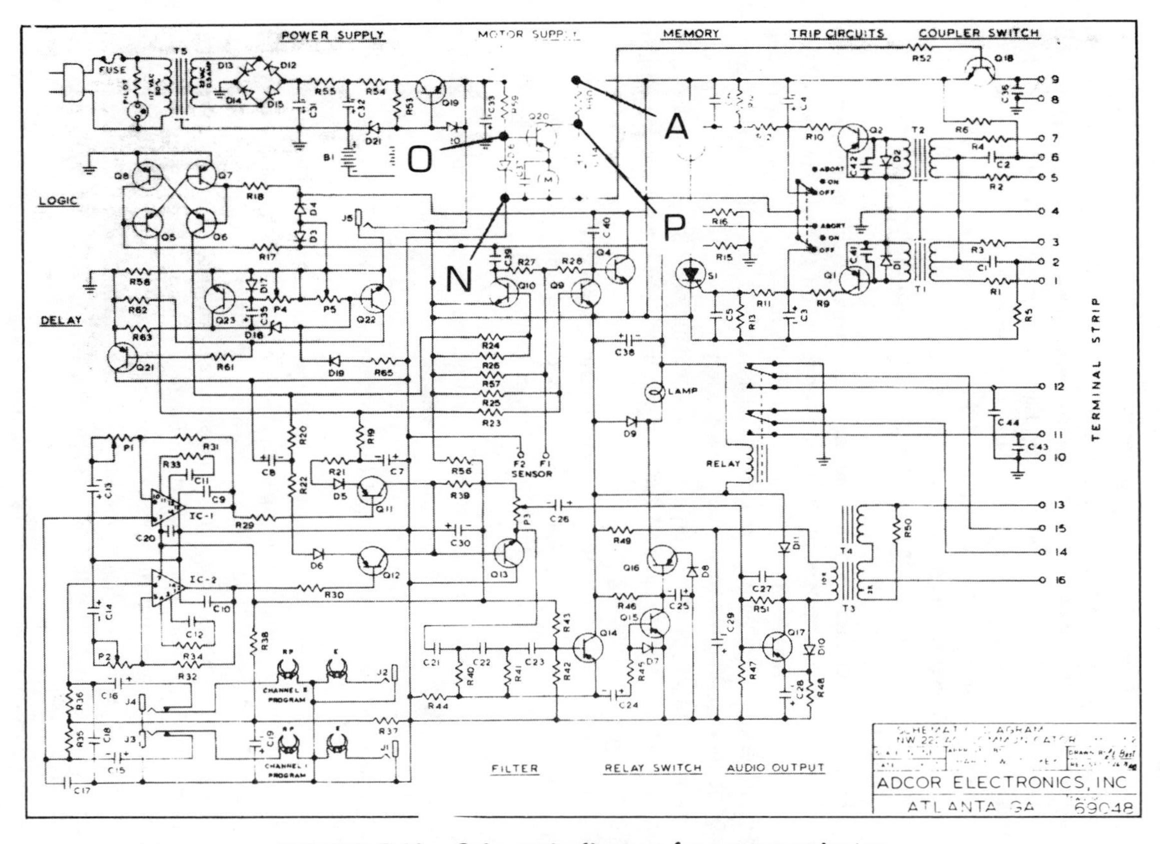

FIGURE 7-12. Schematic diagram for a communicator.

WIRING DIAGRAMS

Complete schematic wiring diagrams are frequently furnished with a security/fire-alarm system to aid in troubleshooting the system, and it is important to have a thorough understanding of this type of drawing.

Components in schematic wiring diagrams are represented by symbols, and every wire is either shown by itself or included in an assembly of several wires which appear as one line on the drawing. Each wire in the assembly, however, is numbered when it enters and it keeps the same number when it emerges to be connected to some electrical component in the system. Figure 7-12 shows a complete schematic wiring diagram for a piece of security equipment. Note that the diagram shows the devices (in symbol form) and indicates the actual connections of all wires between the devices.

Electronic schematic diagrams indicate the scheme of the plan according to which electronic or control components are connected for a specific purpose. Diagrams are not normally drawn to scale, and the symbols rarely look exactly like the component. Lines joining the symbols representing electronic or control components indicate that the components are connected.

To serve all its intended purposes, the schematic diagram must be accurate. Also, it must be understood by all qualified personnel, and it must provide definite information without ambiguity.

The schematics for a control circuit should indicate all circuits in the device. If they are accurate and well prepared, it will be easy to read and follow an entire closed path in each circuit. If there are interconnections, they will be clearly indicated.

In nearly all cases the conductors connecting the electronic symbols will be drawn either horizontally or vertically. Rarely are they slanted.

A dot at the junction of two crossing wires means a connection between the two wires. The absence of a dot in most cases indicates that the wires cross without connecting.

Schematic diagrams are, in effect, shorthand explanations of the manner in which an electronic circuit or group of circuits operates. They make extensive use of symbols and abbreviations. The more commonly used symbols were explained earlier in this chapter. These symbols must be learned to be able to interpret control drawings with the speed required in the field or design department. The use of symbols presumes that the person reading the diagram is reasonably familiar with the operation of the device, and that he or she will be able to assign the correct meaning to the symbols. If the symbols are unusual, a legend is normally provided to clarify matters.

Every component on a complete schematic diagram usually has a number to identify it. Supplementary data about such parts are supplied on the diagram or on an accompanying list in the form of a schedule, which describes the component in detail or refers to a common catalog number familiar in the trade.

To interpret schematic diagrams, remember that each circuit must be complete in itself. Each component should be in a closed loop connected by conductors to a source of electric current such as a transformer or line voltage. There will always be a conducting path leading from the source to the component and a return path leading from the component to the source. The path may consist of one or more conductors. Other components may also be in the same loop or in additional loops branching off to other devices. For each electronic component, it must be possible to trace a completed conducting loop from the souce and back to the source.

8

TROUBLESHOOTING ELECTRICAL/ELECTRONIC ALARM SYSTEMS

A great deal of the work performed by installers of security/fire-alarm systems involves the repair and maintenance of existing security systems. To properly maintain such systems, the workers must have a good knowledge of what is commonly known as *troubleshooting*—the ability to determine the cause of any problem in a security/alarm system and then to correct it in the shortest possible time with the least amount of expense or inconvenience to the building owners.

Troubleshooting covers a wide range of problems from such small jobs as finding a short circuit in a perimeter loop (sensor circuit) to tracing out defects in a complex control circuit. In any case, troubleshooting usually requires only a thorough knowledge of testing equipment and a systematic and methodical approach to the problem; that is, testing one part of the circuit or system after another until the trouble is located.

Those involved in the maintenance of electrical or electronic security/fire-alarm systems should keep in mind that every troubleshooting or electrical/electronics problem can be solved, regardless of its nature. The paragraphs that follow are designed to aid those involved in such work better solve the more common security/fire-alarm system problems in a safe and logical manner.

BASIC ELECTRICAL PROBLEMS

In general, there are only three basic electrical troubles:

1. A short circuit
2. An open circuit
3. A change in electrical value

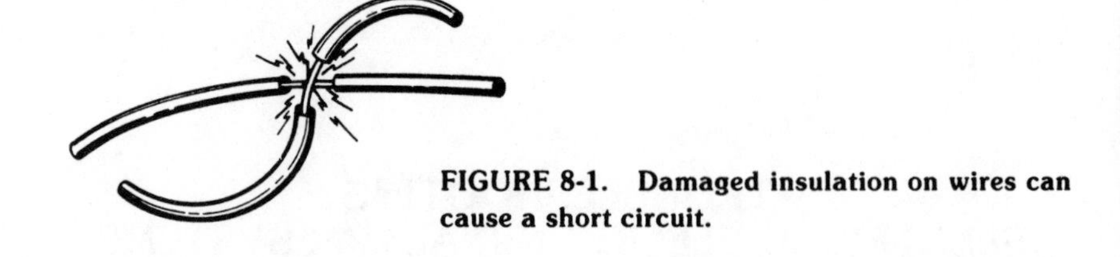

FIGURE 8-1. Damaged insulation on wires can cause a short circuit.

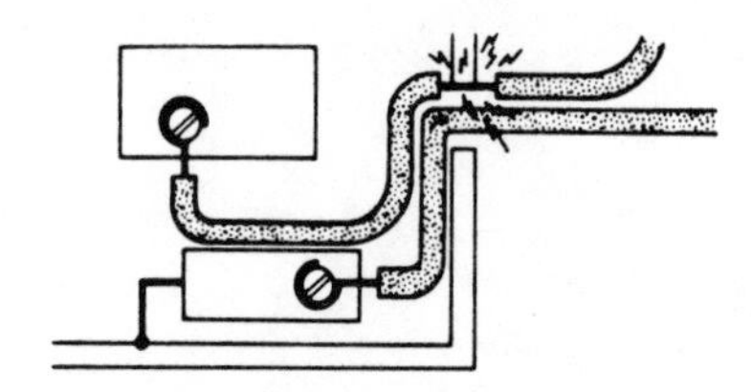

FIGURE 8-2. A faulty "hot" conductor touching a grounded frame can cause a short circuit.

A short circuit is probably the most common cause of electrical problems. Such a condition can be caused by any number of problems, but in most cases the cause is faulty insulation on conductors which allows two conductors to touch (Fig. 8-1) and short out, or else the fault occurs between one "hot" wire and a grounded object as illustrated in Fig. 8-2.

An open circuit is an incomplete current path and is usually caused by either a blown or tripped overcurrent-protection device, a loose connection or broken conductor, or a faulty switch or control.

A change in electric value covers such items as low voltage, electrical surges, a change in resistance, and similar items.

TESTING INSTRUMENTS

To maintain and troubleshoot existing security/fire-alarm systems, workers should know and apply modern testing techniques and have a good understanding of basic testing instruments.

When you use any testing instrument (or meter), always consider your personal safety first. Know the voltage levels and shock hazards of all equipment to be tested and make certain that the instrument used has been tested and calibrated. This should be done at least once a year. To prevent damage to the instrument, select a range (on meters with different ranges) that ensures less than full-scale deflection of the needle. A midscale (or higher) deflection of the needle usually provides the most accurate reading.

Volt-ohm-ammeters The combination volt-ohm-ammeter is probably the technician's most useful testing instrument for electrical systems under 600 V. Many of these instruments are very compact and can easily be carried in a leather pouch attached to the electrician's belt for immediate use.

One type of volt-ohm-ammeter is shown in Fig. 8-3. Instruments of this type are commercially available in current ranges from 6 to over 1000 A and voltages from 0 to 600 V. A separate battery attachment is supplied for use of the ohmmeter scale built into the instrument.

Make sure that the battery-attachment case is removed from the instrument when you are taking current or voltage readings. If the battery attachment is not removed, incorrect voltage or current readings will be obtained. The battery attachment is to be inserted into the instrument only when it is used as an ohmmeter.

To take current readings, release the pointer on the scale by moving the pointer lock button to the left. Turn the scale selector knob until the highest current range appears in the scale window. Press the trigger button to open the jaws before encircling one of the conductors under test with the transformer jaws. *Never* encircle two or more conductors, as shown in Fig. 8-4*a*; only encircle *one* conductor as shown in Fig. 8-4*b*. Release finger pressure on the trigger slowly to allow the jaws to close about the conductor and keep an eye on the scale while doing so. If the pointer jumps abruptly to the upper range of the scale before the jaws are completely closed, the current is probably too high for the scale used. Should this happen, release the jaws immediately and use either a higher scale or a "rating-busting" attachment. If the pointer deflects normally, close the jaws completely and take the reading from the scale.

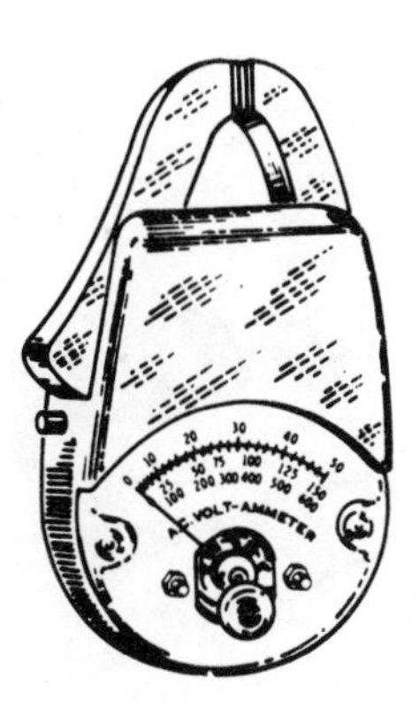

FIGURE 8-3. Typical volt-ohm-ammeter.

FIGURE 8-4. (*a*) Wrong way to make a current reading with transformer jaws. (*b*) Right way to make a current reading.

A reading below half scale indicates that an adjustment is necessary for a more accurate reading. There are two ways to do this. You can increase the accuracy by looping the conductor two or more times around the transformer jaws and then dividing the reading by the number of turns. If the instrument has a lower scale adjustment, it is easier to set the scale at the next lowest range. For example, if the 100-A scale is used and the pointer is below 40 A, as shown in Fig. 8-5*a*, set the rotary scale selector to the next lower current range, and the reading will be in the upper half of the scale, as shown in Fig. 8-5*b*. When very low current readings are encountered, another attachment (Fig. 8-6) is available for increasing

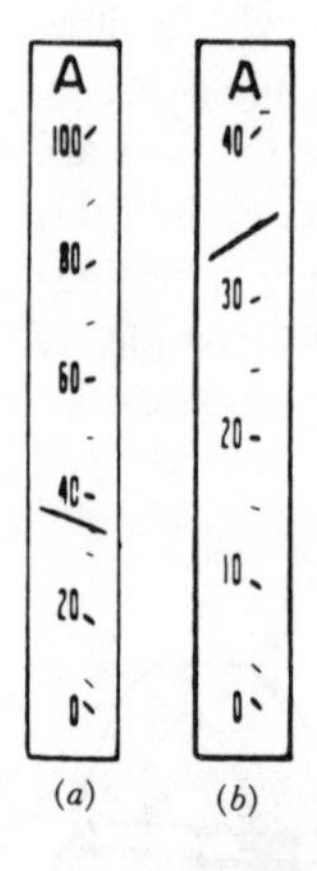

FIGURE 8-5. (*a*) When a 100-A scale is used, the pointer is in the lower half of the scale. (*b*) When a 40-A scale is used, the pointer is in the upper half of the scale, giving a more accurate reading.

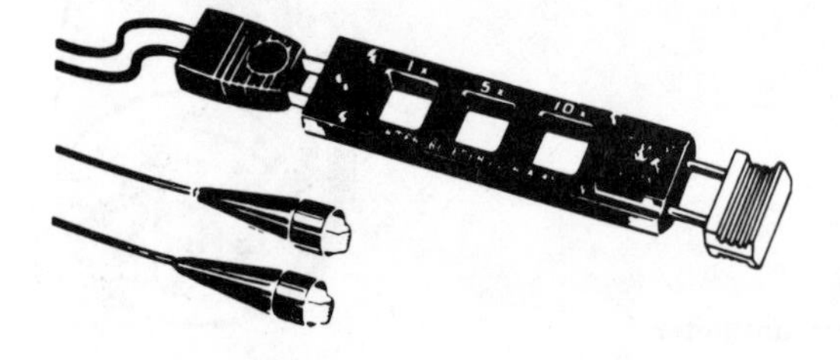

FIGURE 8-6. An adapter for taking very small current readings on a standard ammeter.

the current-measuring sensitivity of the instrument 5 to 10 times. Therefore, the 0- to 6-A range becomes either 0 to 1.2 or 0 to 0.6 A.

HOW TO TAKE VOLTAGE READINGS

Voltage readings are always taken across the circuit (between two phases or one phase to ground) by means of test leads. On the instrument under discussion, the test leads are inserted into the voltage receptacles at the bottom of the instrument. The rotary scale selector is then turned until its highest voltage range, usually 600 V, appears in the scale window.

Connect one alligator clip to one side of the line. Then, with meter in hand, touch the other side of the line with the alligator clip. If the voltage does not exceed 600 V, attach the second alligator clip and read the voltage on the red scale marked 600 V. If the voltage does not exceed 300 V, attach the second alligator clip and read the voltage on the red scale marked 300 V, as shown in Fig. 8-7. If the voltage is below 150 V, rotate the scale selector until the 150-V range appears in the window. Read the voltage on this scale.

To show how to read the scale, let us assume that the pointer is at the position indicated in the scales in Fig. 8-8. In Fig. 8-8*a*, the pointer reads 440 V; each subdivision between 400 and 500 is 20 V. In Fig. 8-8*b*, the pointer reads 78 A; each subdivision between 70 and 80 is 2 A. The heavy mark between 60 and 80 is 70 A. Fig. 8-8*c* shows a reading of 12.7 A. The heavy mark above 12 is 13 A, and each subdivision between 12 and 13 is 0.5 A.

How to use an ohmmeter The instrument under discussion may be changed to operate as an ohmmeter by inserting one of the test leads into the left (with the scale facing you) voltage receptacle at the bottom of the instrument. Insert the other test lead into the bottom of the separate battery case, then push in and lock in place. The opposite end of the battery case is then plugged into the jack on the right-hand side of the instrument just below an ohmmeter zero-adjustment knob.

FIGURE 8-7. Method of using a voltmeter to take a voltage reading between phases.

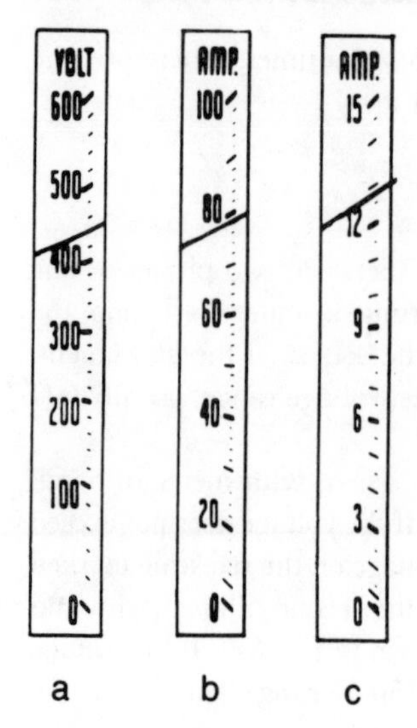

FIGURE 8-8. **(*a*) Pointer reading is at 440 V. (*b*) Pointer reads 78 A. (*c*) Pointer reads 12.7 A.**

Set the range selector so that the 150-V red scale appears in the scale window. The ohmmeter scale should be adjusted with the test leads open or not touching each other. In this position, the pointer should line up with the division marked ∝ (infinity) on the ohm scale. If it does not, turn the pointer zero-adjust screw until it does line up properly.

Now touch the test leads together and the pointer should jump to zero on the scale. If not, use the scale-adjustment knob to line the pointer to zero on the scale.

To measure resistance, make certain that no voltage is present on the device under test and then place the test leads between any two points on which a resistance reading is desired. The ohmmeter scale is located on the flat plate to the right of the window. The zero mark (beginning) is on top of the scale, while the infinity mark ∝ ends the scale.

Many security/fire-alarm systems operate on a dc power supply—involving conventional electronic components. The uses for a good volt-ohmmeter in troubleshooting the circuits are practically endless. Good quality is always a prerequisite for these instruments. If reference resistors of a poor tolerance percentage are used in these devices, readings will not be as accurate as is required for many purposes. Generally, the higher the ohms-to-volt ratio, the more accurate the readings will be. An inexpensive, poor-quality meter can be used to give only a very rough indication of operation and, owing to inaccurate or incorrect readings, may be more of a hindrance than a help to power-supply checkout and operation.

Other electronic measuring devices of limited use in power-supply checking include ac ammeters, which are inductively coupled to the primary ac line to indicate alternating current drain. No physical connection is actually made, and most of these devices simply clip around the ac line cord. Sometimes a capacitance meter is also used to measure the capacitance value of filter components. Specialized power-supply designs may require other types of measuring devices, but for most purposes a volt-ohmmeter will be adequate.

Electronically regulated power supplies using transistors can usually be checked by using the basic volt-ohmmeter circuit, but there are commercially manufactured transistor and diode checkers that will test these solid-state devices in seconds. These instruments can be very expensive if the automatic types are purchased. They require no special test lead connections. The three leads are connected at random to the transistor leads and an immediate indication of the quality of the transistor is given.

Special test equipment such as frequency counters may be required in dc-to-dc power supplies. These devices measure the frequency of the ac portions of the supplies. This is often a necessary measurement for dc-to-dc supplies and for dc-to-ac inverters that must power frequency-sensitive electronic equipment. The volt-ohmmeter is also necessary for the checks that must be made on dc-to-dc power supplies to determine the value of output and input voltage and current.

Safety is a prime consideration when checking out circuits in potentially lethal power supplies. Equipment that may not be operating properly is especially dangerous, because circuit inconsistencies can cause normally safe circuit points to carry a dangerous potential force. Some of the inexpensive volt-ohmmeters may be capable of measuring voltages with values as high as 5000 Vdc, but their test leads may not be insulated to this high value. With inadequate insulation, the danger of a severe electric shock is always a possibility. This is another excellent reason for using only quality test instruments when attempting to determine the operating characteristics of electronic equipment such as dc power supplies.

Another type of testing meter that has gained popularity over the past decade is the digital meter. A digital meter displays the readings (voltage, amperage, ohms, etc.) in digits instead of using a meter movement. See Fig. 8-9. Digital meters have several advantages over testing instruments with the conventional dial indicator. The greatest advantage is that the input impedance, or resistance, is higher. Dial or analog meters normally have a resistance of about 5000 Ω per volt. Therefore on a 3-V full-scale range, the meter movement has a resistance of 15,000 Ω connected in series with it. Consequently, on a 600-V full-scale range, the meter movement has a resistance of 3 million Ω (3 MΩ) connected in series with it.

On the other hand, digital meters commonly have an input impedance of 10 million Ω (10 MΩ), regardless of the range they are set to operate on. The advantage of this high input impedance is that it does not interfere with a low-power circuit. On line voltage readings, the advantage is not very great. However, when working with the lower voltage control circuits for security/fire-alarm systems, a low impedance meter will not only lead to inaccurate readings, but it may also alter the circuits themselves.

Another advantage of the digital meter is that it is generally easier for an inex-

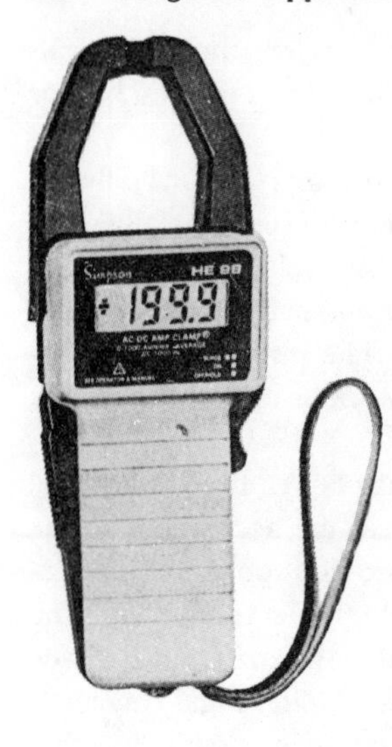

FIGURE 8-9. Simpson Electric Co. clamp-on digital ammeter. Digital multimeters are also available for taking voltage, amperage, and resistance readings.

perienced person to learn to read. Dial or analog meters can be used for the majority of testing and troubleshooting when working with security/fire-alarm systems, but it usually takes more experience and practice for the technician to read the analog meters—especially on the lower values.

PRACTICAL APPLICATION

Most manufacturers of security/fire-alarm systems publish service manuals for their equipment which present—in a simplified and systematic manner—troubleshooting and servicing procedures for their equipment. Their main objective is to direct a security/fire-alarm technician to the exact cause of a malfunction and assist in repairing the system. Even qualified technicians not familiar with the system should be able to isolate the exact cause of nearly all possible malfunctions when using such manuals. Therefore, it is recommended that all service technicians obtain a service manual (including wiring diagrams, etc.) for the particular system in question.

Although the exact content of service manuals will vary from manufacturer to manufacturer, most are divided into three main sections:

1. Theory of operation
2. Troubleshooting
3. Repair and adjustments

The section on theory of operation should give a thorough explanation of each circuit and should be accompanied by a schematic diagram—such as the one in Fig. 8-10—clearly identifying points that are referred to in the text. The purpose

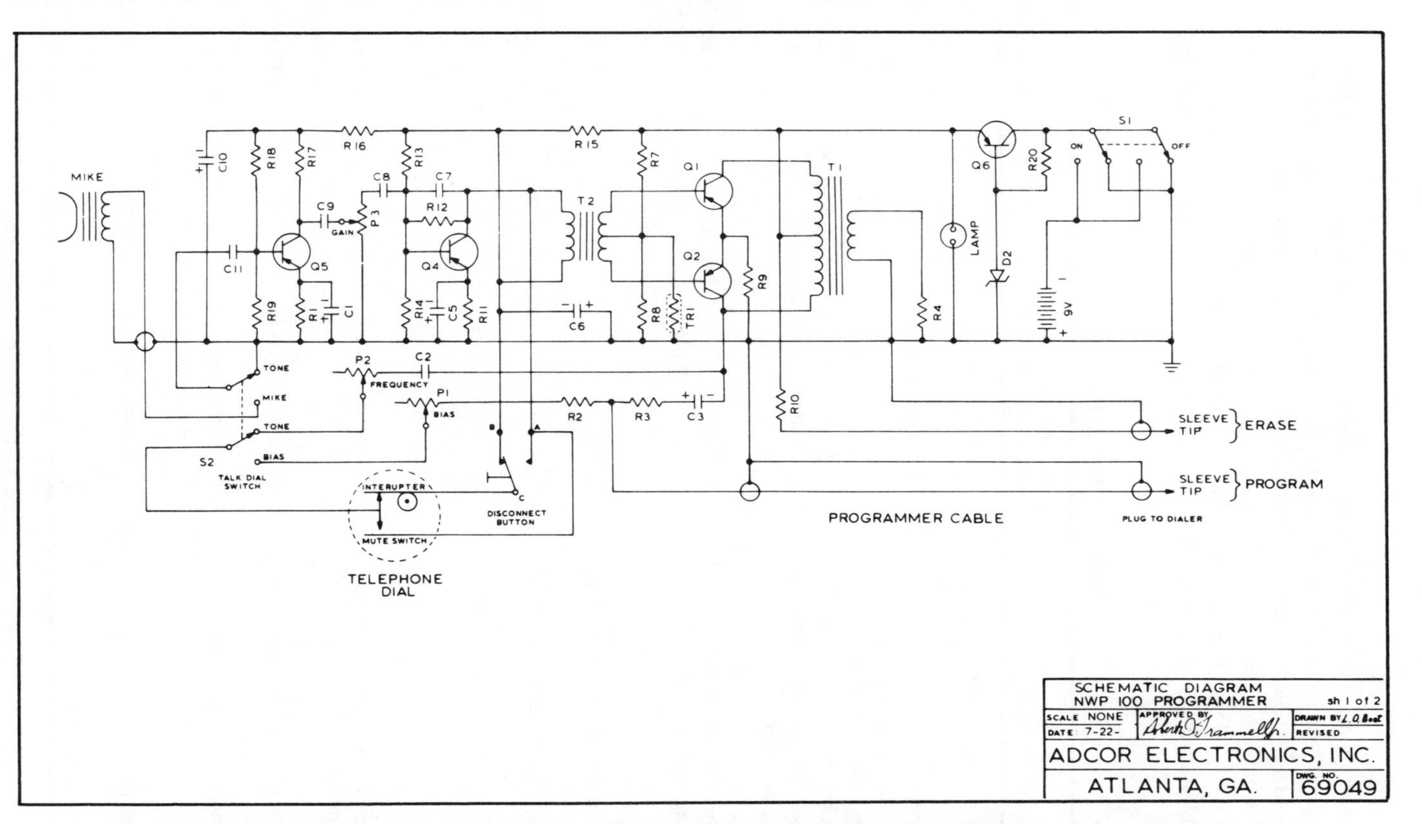

FIGURE 8-10. Schematic diagram accompanying the "Theory of Operation" section of a service manual.

of this section is to give technicians an understanding of the proper operation of the system, which in turn will enable them to locate the causes of those problems not found by following the regular step-by-step troubleshooting procedures.

A troubleshooting table such as the one in Fig. 8-11 lists the symptoms that may occur, along with the most likely causes and page numbers of the detailed troubleshooting procedures for the various circuits. The service technician will find

SYSTEM TROUBLE-SHOOTING

Symptom	***Possible causes*** *(check in order listed)*	*Reference*
Resistors R54, R55 burnt	1. Power supply	1. Page 4B
Dialer does not trip Note: Make sure switch is in ON (center) position. Make sure potentiometer P4 is in extreme clockwise position as viewed from terminal strip end of alarm center.	1. Batteries weak 2. Power supply 3. Trip circuits 4. Start-delay and battery cutout circuit 5. Channel switch and logic	1. Page 4B 2. Page 4B 3. Page 5B 4. Page 7B 5. Page9B
Motor does not run on both channels and lamp does not light	1. Start-delay and battery cutout 2. Motor supply 3. Channel switch and logic	1. Page 7B 2. Page 8B 3. Page 9B
Motor does not run but lamp comes on	1. Motor supply	1. Page 8B
Motor runs but lamp does not light	1. Dialing filter and relay control	1. Page 12B
Dialer trips but does not shut off	1. Memory circuit 2. Trip circuit 3. Channel switch and logic	1. Page 6B 2. Page 5B 3. Page 9B
Dialer cannot switch from one channel to the other	1. Memory circuit	1. Page 6B
Dialer cannot program, erase	1. Tape head, jacks	1. Page 10B
Audio output weak, distorted	1. Output amplifier 2. Audio amplifier	1. Page 11B 2. Page 10B
No dialing pulses	1. Dialing filter and relay control	1. Page 12B
Coupler not functioning	1. Coupler switch	1. Page 13B

FIGURE 8-11. Typical troubleshooting table.

that consulting this table, and checking the *possible causes* in the order given, will prove to be the most efficient procedure to follow. The possible causes are listed either in order of most probable cause or fastest to verify cause, whichever has been found by the manufacturer's technicians to be the most efficient approach.

A repair and adjustment section usually is included to provide guidance in replacing defective components, or realigning components that have gotten out of adjustment. When such instructions are given, the service technician should follow them closely to ensure successful repair. Of course, technicians performing the repair must have a working knowledge of repair procedures and precautions that pertain to printed-circuit boards. They should also be equipped with the proper tools. Lack of ability and/or use of improper tools can lead to more damage rather than eliminate a problem.

There is really no substitute for the service manuals provided by the manufacturers, and all service technicians should obtain manuals for the equipment on which they are working as well as for equipment that they expect to be preforming work on.

9

ESTIMATING AND CONTRACTING FACILITIES

Persons engaged in the installation of security/fire-alarm systems know that the only way to continue profitable operations is to practice extremely efficient management techniques, to use only workers specifically trained or experienced and adapted to the work, and to develop an accurate and relatively fast system of estimating the cost of the installation. A review of basic estimating procedures as they relate to security/fire-alarm installations, together with an example, will serve to point out problems and procedures peculiar to this type of work.

OFFICE LOCATION AND LAYOUT

The physical arrangement of the shop and office is an important factor in providing for the servicing of the work at the lowest possible cost and making best use of materials and tools.

Sometimes, because of the limitations of type of building, type of building access, floor area, relation to streets, parking, and so on, it may not always be possible to provide an ideal physical arrangement of facilities. However, intelligent planning can usually provide for maximum efficiency under the given conditions.

Ideally, the contractor's shop should provide for:

1. Ease of handling materials and tools with a minimum of shifting around and extra handling

2. Ready accessibility

3. Sufficient predelivery storage or holding areas for assembled orders for specific jobs and holding areas for materials and tools returned from jobs awaiting checking and return to inventory

4. Unobstructed truck access

In general, these objectives are best accomplished when (1) all facilities are located on one floor; (2) adequate floor area is available; (3) there are sufficient bins, shelving, and storage racks with adequate aisles and walkspace; and (4) driveway access from the street to the building can be made from streets that are not congested with traffic.

When the available floor area is restricted, it is often possible to construct a balcony which will increase the total floor area and which can also be used for the storage of items which are handled and shipped less frequently.

OFFICE EQUIPMENT

The type and amount of office equipment will naturally vary depending upon the volume of work performed by the contractor. For a one-person operation, however, the following should be considered minimum:

1 owner's desk and chair

1 secretary's desk and chair

1 four-drawer file cabinet

1 plan rack

1 typewriter

1 bookcase (catalogs)

1 drawing table

2 chairs for customers

Of course, where bidding work is necessary (and it will be), a complete estimating facility should be provided, but these facilities are described later in the chapter. This is a rather skimpy list, but a new contractor can get by with these items until profits warrant buying additional equipment. Of course, the contractor will need other miscellaneous items such as letterheads, envelopes, estimating and possibly drafting tools, bookkeeping supplies, an electronic calculator, and so on.

The cost of a setup for a one-person operation will be in the neighborhood of $5000.

ESTIMATING FACILITIES

Estimating facilities in a contractor's office will vary in size, but a separate area should be provided for the purpose so as to be out of the way of normal office routines. It should be illuminated with 100 to 150 fc of well-diffused light, painted a pleasing color (preferably a pastel shade), provided with adequate heating, cool-

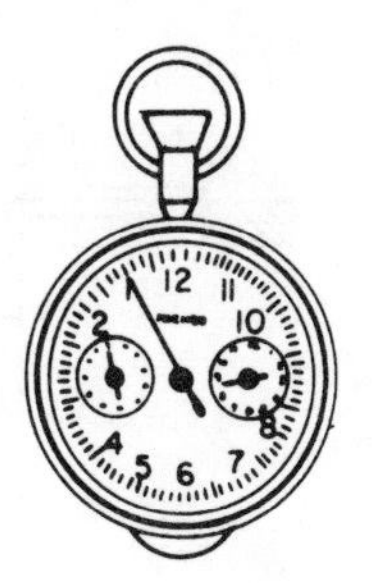

FIGURE 9-1. A rotometer is used to measure conductor lengths on scaled drawings—especially where curved lines are encountered.

ing, and ventilation, and of sufficient size to accommodate all necessary estimating tools, materials, and drawings.

The furnishings within the area should consist of a drawing board, utility or throwoff table or desk, comfortable seats, filing cabinets, and plan files. Bookshelves for manufacturers' catalogs, estimating manuals, and the pricing data should be placed near the work area for easy access.

The estimator should be equipped with an electronic calculator, drafting instruments and supplies, a rotometer (Fig. 9-1), a metal tape graduated in various scales (Fig. 9-2), colored pencils, and a tabulator.

System components, such as outlets and door and window contacts, are counted with the tabulator. Circuit conductor lengths are measured with the rotometer on scaled drawings. If scaled elevation drawings are available, the rotometer may also be used to measure the lengths of window foil needed. The colored pencils are helpful in checking off components and wiring as they are accounted for.

Two drafting facilities are shown in Fig. 9-3 and 9-4; the first would be used by a one-person firm, the second by a larger firm using several estimators, designers, and drafters. A study of these illustrations should help the reader decide upon a suitable layout for his or her own estimating facilities.

Take, for example, the floor plan in Fig. 9-4. This estimating room is large enough to accommodate four estimators by using U-shaped counters along one entire wall; this gives three of the estimators an L-shaped work area. Drawing

FIGURE 9-2. A metal tape, graduated in various scales, is useful for measuring long conductor runs on drawings.

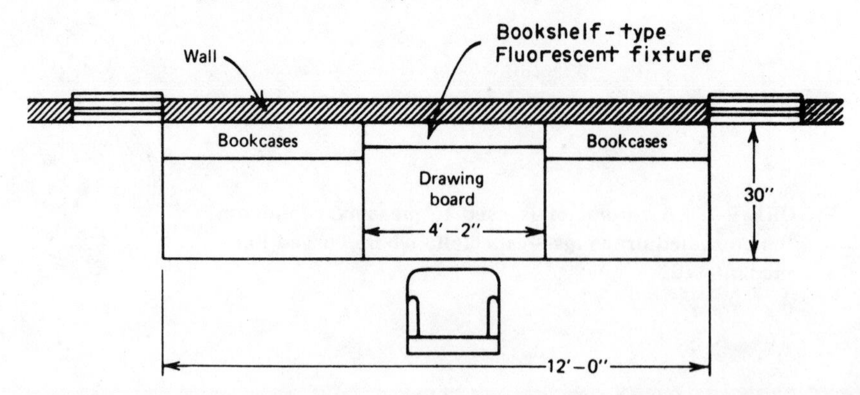

FIGURE 9-3. Drafting/estimating facilities for a one-person firm.

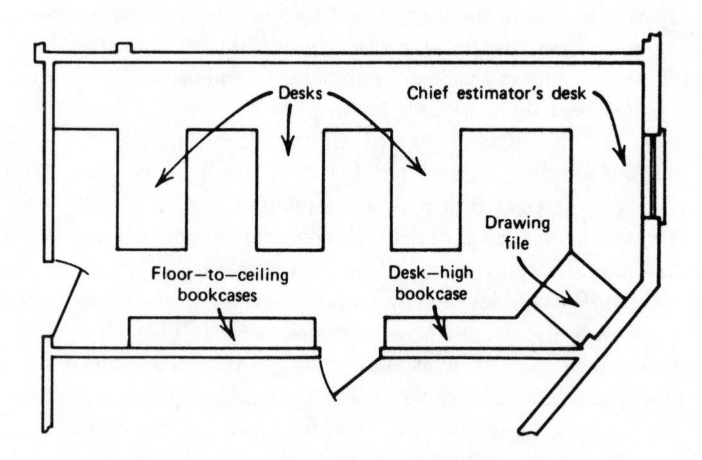

FIGURE 9-4. Estimating facilities for four workers.

boards are mounted on one leg of the L-shaped counter tops while the remaining space is used for reference materials or throwoff space.

Duplex receptacles are located above the counter tops along the wall to provide electricity for electrically operated devices such as electric erasers and calculating machines.

The wall opposite the counters contains bookshelves for storage of manufacturers' catalogs and equipment data. At the end of each counter (containing the drawing boards) is a built-in bookcase the same width and height as the counter itself. These bookcases contain reference books most frequently used by the estimator at each space—estimating manuals, design data, and so on.

Notice that the end area next to the window is somewhat larger than the three remaining areas. This area was planned for the chief estimator, who handles more drawings and must have more work space to organize the estimating procedures. The plan file, to the chief's immediate right, contains drawings of projects under construction. Thus, these drawings are at hand when questions are phoned in by the workers on the job.

To the immediate left of the drawing board is a large throwoff space with a plan file. The walls above the counter contain charts in large print of the most-used estimating and design data for quick reference. Two-drawer legal-size filing cabinets installed under the counter tops are used for storage of job files and other reference material.

The chief estimator is also provided with a telephone, a tape recorder for dictation, an electronic calculator, and an assortment of other drafting and estimating tools and materials.

ESTIMATING FORMS

Many calculations and records must be made during the process of estimating security/fire-alarm systems. Such calculations are best performed when a systematic pattern is followed, using appropriate forms and takeoff sheets. In general, all forms should have spaces for the name of the project, the date, the name of the estimator, the name of the owner, and other standard data. The following forms will be discussed:

1. Estimate sheet (Fig. 9-5)
2. Bid summary sheet (Fig. 9-6)
3. Job sheet (Fig. 9-7)
4. Job record insert (Fig. 9-8)
5. Pricing sheet (Fig. 9-9)
6. Job progress report (Fig. 9-10)
7. Small takeoff and listing sheet (Fig. 9-11)
8. Large takeoff and listing sheet (Fig. 9-12)
9. Bid and estimate summary (Fig. 9-13)
10. Alarm system proposal (Fig. 9-14)

Forms 1 through 9 are available from the Minnesota Electrical Association, 3100 Humboldt Ave. South, Minneapolis, MN 55408. The alarm-system proposal form may be obtained from Conrac Corp., Old Saybrook, CT 06475.

The estimate sheet is arranged with description, material, and labor columns for orderly recording of costs. The use of separate sheets is suggested for each floor or section of a project.

ESTIMATE SHEET

JOB ______ PAGE ______

______ OF ______ PAGES

ESTIMATED BY ______ CHECKED BY ______ DATE ______

DESCRIPTION	MATERIAL				LABOR		
	Quantity	Unit Price	Per	Amount	Unit	Per	Amount
		TOTAL					

MISCELLANEOUS		RECAPITULATION	
		Material Cost	
		Hours Labor @	
		Hours Labor @	
		Direct Job Expense	
		Total Prime Cost	
		Overhead Expense	
		TOTAL COST	
		Profit	
		Selling Price	
Form E-3		BID SUMMITTED	

FIGURE 9-5. Estimate sheet.

The bid summary sheet has a complete checklist to help the contractor avoid some of the common errors made in estimating security/fire-alarm systems.

The job sheet has spaces for a complete cost record of the average job. The front provides for labor records, job expense, tools and equipment charges, cost summary, and all other necessary information. The reverse side provides space for listing materials taken out, amount returned, net quantity used, cost, and selling

BID SUMMARY SHEET

JOB ______________________________ SHEET NO. ___ OF ___ SHEETS

ESTIMATED BY ____________ CHECKED BY ____________ DATE ____________

SHEET NO	DIVISION	MATERIAL DOLLARS	LABOR HOURS

NON-PRODUCTIVE LABOR	HOURS
Handling Material	
Superintendent	
Traveling Time and Lost Time	
Job Clerk	
TOTAL (A)	

JOB EXPENSE	DOLLARS
Tools, Scaffolds	
Pro Rata Charges	
Insurance, Public Liability, Etc.	
Cutting, Patching, Painting	
Watchman	
Telephone	
Drawings	
Inspection and Permit Fees	
License	
Storage	
Freight, Express and Cartage	
Transportation	
Board ___ Men ___ Weeks At ___	
TOTAL (B)	

	MATERIAL DOLLARS	LABOR HOURS
Miscellaneous Material and Labor		
Non-Productive Labor		(A)
TOTALS MATERIAL (C) & LABOR (D)		
(D) ___ Hours Labor @ ___		
___ Hours Labor @ ___		
___ Hours Labor @ ___		
Taxes: Soc. Sec. ___ Unemp. ___		
Workmen's Compensation Insurance		
LABOR COST GROSS TOTAL		
Job Expense (B)		
Material Cost (C)		
TOTAL PRIME COST		
___ % Overhead		
TOTAL NET COST		
___ % Profit		
Selling Price Without Bond		
Bond		
Selling Price With Bond		
PRICE QUOTED		

To Avoid Errors, Check List on Reverse Side   Let's Upgrade Our Electrical Industry

FIGURE 9-6. Bid summary sheet.

prices. This form is indispensable for accurate and precise control over material and job costs. While this form is not used during the normal estimating process, it is helpful for comparing costs of previous jobs when a similar new job is being estimated.

The job record insert is designed for the workmen to take to the job and return later for filing. Space for pertinent information is provided on the front with col-

JOB SHEET

PERMIT NUMBER ____________ JOB NUMBER ____________

NAME ____________ ADDRESS ____________

JOB AT ____________ KIND OF JOB ____________

START ______ FINISH ______ RECAP ______ INVOICE ______ CUSTOMER LEDGER ______

ORIGINAL PRICE			BILLING PRICE		
EXTRAS			TOTAL COST (See Below)		
TOTAL BILLING PRICE OF JOB			NET PROFIT OR LOSS		

DIRECT JOB EXPENSE			COST SUMMARY		
PERMIT AND INSPECTION			ITEM	Estimated	Actual
COMPENSATION INSURANCE			MATERIAL (See Other Side)		
FINANCING			LABOR (______ Hours)		
COMMISSIONS			DIRECT JOB EXPENSE		
CAR FARE			PRIME COST		
DRAYAGE			OVERHEAD EXPENSE		
			TOTAL COST OF JOB		
			Remarks		
TOTAL JOB EXPENSE					

LABOR COST

Dates	Workman	Hours	Rate	Cost	Rate	Charge
			TOTAL			

TOOLS AND EQUIPMENT

Quantity	Item	Workman	Returned	Used	Charge	Remarks

FIGURE 9-7. Front of job sheet.

umns on the interior of the folder for listing materials used. The estimator can draw a sketch of the proposal directly on the form for use by the workers. The back of the page has space for extra work requested by the customer and room for the customer's signature.

The pricing sheet provides for material and labor listings as well as material

CHECKED BY: __________ JOB NO. __________

OTHER HELP: __________ DATE: __________

OWNER: __________ TYPE OF JOB: __________

JOB ADDRESS: __________ JOB COMPLETED ☐ NOT COMPLETED ☐

PERMIT YES____ NO____

INSPECTOR __________

PHONE __________

INSPECTION CALLED __________

WORK TO BE DONE:

OPINION OF JOB: __________

CUSTOMER'S ACKNOWLEDGEMENT:

This work has been satisfactorily completed:

SIGNED: __________ (Customer)

E-22

JOB RECORD INSERT

FIGURE 9-8. Job record insert.

prices and labor units. Separate pricing sheets should be used for each floor or section of the project and summarized on the bid summary described above.

A material and price sheet is another type of summary sheet that can save the contractor much valuable time in listing each item. This form provides spaces for all commonly used materials, quantities consumed on the job, and the pricing of

PRICING SHEET

JOB ______________________________ ESTIMATE NO. ________

WORK ______________________________ SHEET NO. ____ OF ____ SHEETS

ESTIMATED BY ________ PRICED BY ________ EXTENDED BY ________ CHECKED BY ________ DATE ________

	MATERIAL	MATERIAL					LABOR			
		QUANTITY	LIST PRICE	PER	DISC.	EXTENSION	UNIT	PER	EXTENSION	
1										1
2										2
3										3
4										4
5										5
6										6
7										7
8										8
9										9
10										10
11										11
12										12
13										13
14										14
15										15
16										16
17										17
18										18
19										19
20										20
21										21
22										22
23										23
24										24
25										25
26										26
27										27
28										28
29										29
30										30
31										31
32										32
33										33
34										34
35										35
	FORM E-12									

FIGURE 9-9. Pricing sheet.

these items. This form is particularly adaptable to residential and small commercial projects, although it will serve most general estimating purposes.

The job progress report is more of a management tool than one for estimating. At a glance, it shows how much time has been used in hours and dollars. The same is true for all materials and other direct job expenses, including costs of

JOB PROGRESS REPORT

JOB ______ Contract Amount ______ Sheet No. ______ of ______ Sheets

JOB NO. ______ Change Orders ______ Date ______

Dating	MATERIAL		LABOR		DIRECT JOB COST				MISCELLANEOUS		TOTAL COSTS		MAN HOURS	
	Monthly	Total	Monthly	Total	Monthly	Total	Monthly	Total	Monthly	Total	Monthly	Total	Monthly	Total

CODE 1. INSURANCE 2. BONDS 3. PERMITS 4. TOOL COST 5. TRANSPORTATION 6. ______

FIGURE 9-10. Job progress report.

inspection, equipment, hauling, insurance, etc. It is set up on a running-account basis, and a quick comparison is available on this one sheet with the original estimate to show how the job is progressing and how much material and what labor costs have not been used. It is a great help in preparing bids for similar jobs while this one is in progress.

The small takeoff and listing sheet is used to list the length of circuit runs, outlets, etc., as they are measured and counted from a drawing. They must be listed first to obtain total quantities of each item or type of material before being transferred to regular estimate or pricing sheets. This form provides an orderly

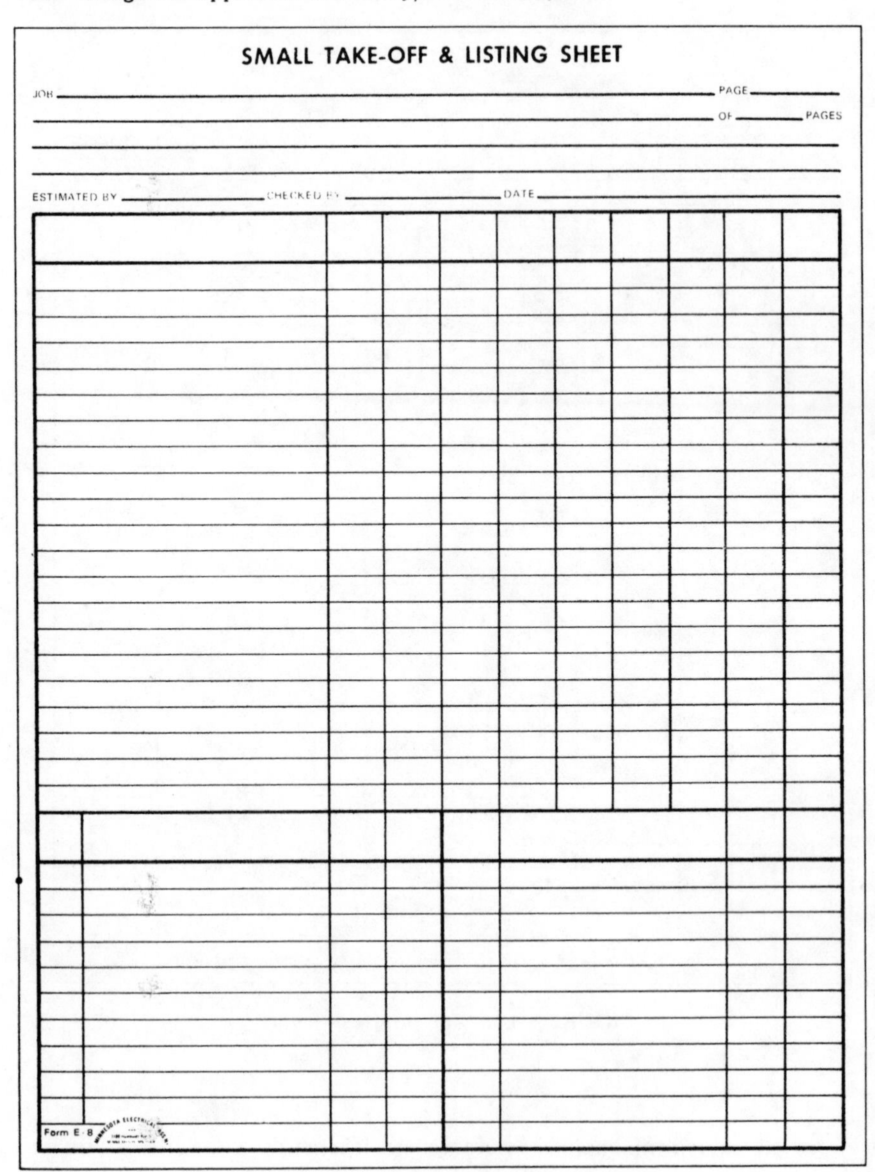

SMALL TAKE-OFF & LISTING SHEET

JOB ______ PAGE ______

______ OF ______ PAGES

ESTIMATED BY ______ CHECKED BY ______ DATE ______

Form E-8

FIGURE 9-11. Small takeoff and listing sheet.

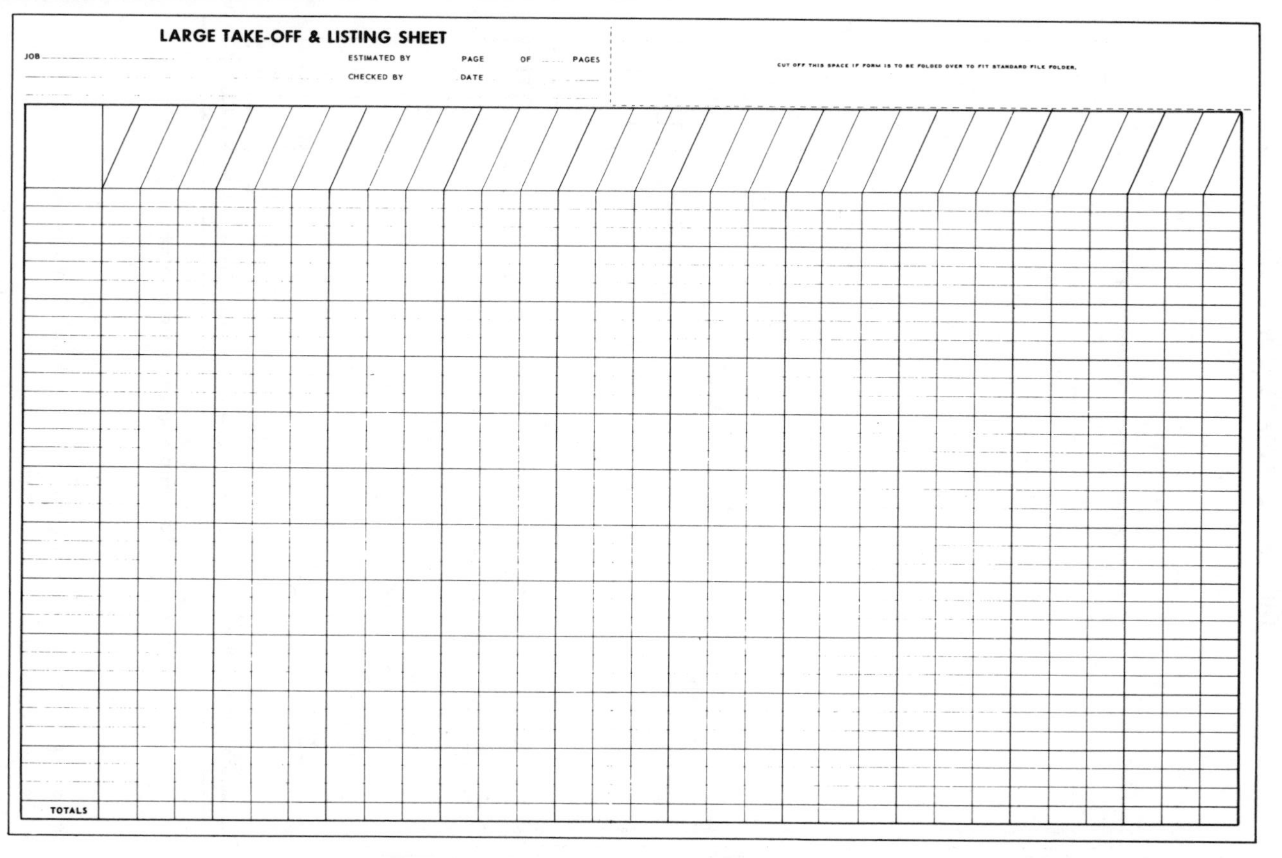

LARGE TAKE-OFF & LISTING SHEET

JOB

ESTIMATED BY

PAGE OF PAGES

CHECKED BY

DATE

CUT OFF THIS SPACE IF FORM IS TO BE FOLDED OVER TO FIT STANDARD FILE FOLDER.

TOTALS

FIGURE 9-12. Large takeoff and listing sheet.

BID AND ESTIMATE SUMMARY

PROJECT NAME________ ESTIMATED BY________ ESTIMATE NO.________

LOCATION________ CHECKED BY________ SHEET NO.____OF____SHEETS

________ APPROVED BY________ DATE________

Form E-6

SCHEDULE I – SUMMARY OF ESTIMATE SHEETS

SHEET NUMBER	SECTION	MATERIAL	LABOR IN HOURS OR DOLLARS
	TOTALS		

SCHEDULE II – LABOR COSTS BY DOLLARS OR MAN-HOURS

(USE LIGHT AREAS FOR FIGURES)

Total Labor from Schedule I		
Labor Job Factor using Percent of Total Labor in Schedule I:		
Weather @ __ % (0-20%)		
Size of Job @ __ % (0-30%)		
Coordination @ __ % (0-10%)		
Complexity @ __ % (0-15%)		
Labor Efficiency Factor @ __ % (± ?%)		
Stand-By (Lump Sum)		
Total Job Factor		
Non-Productive Labor:		
Supervision		
Timekeeper-Stockman-Job Clerk		
Ordering and/or Handling Material		
Lost Time and/or Traveling Time		
Testing if Required or Desirable		
Other –		
Total Non-Productive Labor		
Total Labor in Man-Hours or Dollars		
(If labor has been figured in man-hours, multiply total above by average labor rate per hour $______)		

FIGURE 9-13. Bid and estimate summary.

means of performing this operation. After the required materials are listed, totals can be transferred to the estimate sheet.

The large takeoff and listing sheet is used for the listing of various materials as they are measured or counted from plans, using one line for each circuit or each room as desired. The various items of material—wire (by sizes), outlet boxes,

ALARM SYSTEM PROPOSAL

FOR: Name ______________________ Date ____________

Street ______________________ City ______________ Phone ____________

LOCATION

PREPARED BY:

NAME ______________________

ADDRESS ______________________

CITY __________ STATE ________ ZIP ________

PHONE ______________________

NO. OF ITEMS PER LOCATION	ITEM	TOTAL QTY.	MODEL NO.	UNIT PRICE	TOTAL PRICE
	CONTROL PANEL				
	POWER SUPPLY				
	EXIT/ENTRY DELAY				
	INDOOR REMOTE STATION				
	OUTDOOR REMOTE STATION				
	PREALARM STATION				
	OUTDOOR SOUNDING DEVICE				
	INDOOR SOUNDING DEVICE				
	DIALER				
	POWER SUPPLY				
	SWITCHING MODULE				
	CONNECTOR CORD				
	TAPE CARTRIDGE				
	ULTRASONIC DETECTOR				
	PHOTOELECTRIC SYSTEM				
	SMOKE DETECTORS				
	HEAT SENSORS				
	MAGNETIC CONTACTS				
	GLASS BREAK DETECTORS				
	SWITCH MAT				
	MISC. WIRING SUPPLIES				

TOTAL NET PRICE	
INSTALLATION	
SALES TAX	
TOTAL PRICE	
DEPOSIT	
BALANCE	

CONRAC
FIRE/LIFE SAFETY AND INTRUSION PRODUCTS

CONRAC CORPORATION • OLD SAYBROOK, CT 06475 • (800) 243-0480

FIGURE 9-14. Alarm-system proposal.

wiring contacts, etc.—are entered, together with the quantities of each. Listing space is provided for 33 items and 30 rooms or sections.

The bid and estimate summary sheet is recommended for contractors who want detailed information included in the sheet format, eliminating the necessity of writing each item separately. The front of the form is used to summarize all costs in making the bid; that is, the first half provides space to list each separate estimate sheet with the total cost of material and labor. When columns for material and labor costs are totaled, the next calculation will be nonproductive labor and job expense. These calculations may be simplified by using columns located at the bottom of the sheet, taking into consideration all costs and profit to arrive at a total price. The reverse side of the sheet has spaces for direct job expense and other pertinent information.

The alarm-system proposal is intended to be presented to the customer when the detailed estimate has been completed. Each room or area of the building is listed in the columns marked "Location," and then the number of items per location is entered in the squares below, "No. of Items Per Location." These items are totaled and priced. The total net price of material, labor costs, sales tax, etc., are totaled to obtain a selling price.

ESTIMATING THE COST OF SECURITY/FIRE-ALARM SYSTEMS

Sound cost estimating of security/fire-alarm systems consists of a complete takeoff (or quantity survey) of all materials and equipment required for a complete installation, and then the calculation of the total labor required to install equipment and materials. To the cost of materials, equipment, and labor are added all direct expenses, variable job factors, taxes, overhead, and profit to determine a "selling price" for the project. If these procedures are intelligently performed and combined with good job management, the estimate should compare very favorably with the actual construction cost.

The steps necessary to prepare a cost estimate for a given electrical installation will normally run as follows:

Takeoff: The count of all lighting fixtures, outlets, and similar items, and the measurement of all branch circuit wiring, feeders, service raceways, etc.

Listing the Material: All items accounted for in the takeoff should be listed in an orderly sequence on a standard pricing sheet, as will be described later.

Applying Labor Units: Determining the proper labor unit from proven labor-unit tables and applying them to the various materials or labor operation under the labor-unit column on the pricing sheet.

Finalizing: The summation of material dollars and labor hours, the assignment of job factors and dollar values to labor hours, and the determination of overhead and profit.

MATERIAL TAKEOFF

A quantity survey or material takeoff consists of counting all the outlets and components by type (door contacts, bells, lockswitches, alarm panels, telephone dialers, etc.). These quantities are entered in their appropriate space on a material takeoff form such as those discussed in Chap 9.

Some estimators make a very detailed material takeoff, listing all circuits separately and including such small items as wire connectors, fastenings, etc. Others will take off the major items of material for an entire building, listing only the different types of materials separately and giving a lump-sum dollar value for small incidentals. It really doesn't matter which procedure is used, as long as the estimator has sufficient information from which to make a complete list of all materials required to complete the installation. With this list it will be possible to apply labor units and thus have a means for pricing and ordering the materials.

A typical material takeoff begins with the counting of all magnetic contacts, glass-break detectors, smoke detectors, heat sensors, sounding devices, etc. The estimator continues by counting all alarm panels, telephone dialers, and all other major equipment. With this material out of the way, the measuring of all wiring, window foil, etc., takes place, using a rotometer or architect's scale.

The actual mechanics of a material takeoff procedure is comparatively simple and will become almost routine in a very short time. The sooner the estimating procedures become routine, the sooner the estimator will be able to make rapid and accurate takeoffs.

The estimator should remember that the circuit lines on floor plans represent only the horizontal portion of the various runs. To accumulate the vertical runs in the system, a scaled section of the building, showing the various floors and ceilings, should be used. Then the mounting heights of alarm stations, sounding devices, etc., can be marked to scale on the drawing of the building section. During the measuring process, whenever a point is reached where there is a vertical section of the run, the rotometer or other measuring device is run over the proper vertical distance on the scaled cross section. This will continually accumulate the vertical distances along with the horizontal distances on the floor plans.

There are several other aids that will help the estimator make an accurate takeoff. One is to use different colored pencils in checking off runs of conduit as they are measured—a different color for each wire size or section of the runs. For example, black could be used to indicate the detection circuits, blue for control circuits, and red for annunciation circuits.

LISTING THE MATERIAL

While taking off the various alarm components from drawings, the estimator must list the items on pricing sheet forms so that pricing may be obtained for the various items and labor units added and extended. To help make this operation easier for both the estimator and the purchasing agent, the listings should be made in an orderly sequence on the pricing sheets.

When the takeoff has been done properly, the estimator will immediately have two items of valuable information: a brief description of each outlet, component, circuit run, equipment, etc., and the quantity of each item listed. From these descriptions, the estimator can determine the exact quantity of materials and the necessary labor hours to completely install the system, provided that he or she has a good knowledge of actual security/fire-alarm installations and building construction.

Such incidentals as fastenings, hangers, wire connectors, etc., are rarely noted on drawings. Therefore, the estimator will be required to make an educated guess, based upon an understanding of the project's requirements, past experience with other projects, and the use of good judgment.

There is really no easy way to accomplish an exact material takeoff, but experienced estimators are able to produce very accurate estimates, rarely omitting important items, when a systematic method is used. The extent of the detail in which items of material are listed can vary to fit the contractor's particular method, but experience has shown that the more detailed the list, the better.

APPLYING LABOR UNITS AND PRICING MATERIAL

Determining the amount of labor that will be expended on a particular project is much more involved than the mere application of labor units. The pricing of materials also requires an intelligent analysis of the quotations by suppliers or price services. The majority of estimators obtain firm quotations from one or more electric equipment suppliers, including the manufacturers of security/fire-alarm equipment.

Requests for quotations on special materials should be made as early as possible, although electric equipment suppliers make it a habit to wait until the last few minutes prior to bid openings before giving out the quotation. This, of course, is done to prevent underbidding by another supplier.

When the quotation is received, the contractor should check over the list of items on the quotation carefully. Electrical suppliers do not normally guarantee that the items will meet with the project's specifications, nor will they take any responsibility for errors. Substitutions are common these days, and it is the contractor's responsibility to make certain that all items quoted will meet with the architect's or engineer's specifications. The contractor should also check the quantities of the quotation against those obtained from the takeoff to make sure they are correct.

Whenever possible, the contractor should obtain a guarantee of the quoted price for a definite period of time. Most suppliers will stand by their quotations for approximately 30 days. But what happens if it takes six weeks to award the contract? There is a good chance that the material quote will go up, and the contractor will have to pay more for the material than the price used in estimating. Therefore, the contractor should try to determine exactly when a particular job will be awarded (This is not necessarily the date of the bid opening.) and then obtain a guarantee that the quote will be good until at least that time.

While waiting for a quotation from suppliers, labor units should be applied to the takeoff. A labor unit is a time figure indicating the time required to install, connect, or otherwise make usable a given item of material or a given labor operation. These units are used by the majority of contractors who must quote a firm lump-sum price to obtain security/fire-systems work. The units are normally based upon worker-hours or a percentage of a worker-hour. For example, 1.50 worker-hours indicates 1½ worker-hours; that is, the labor required for a particular operation will take one worker 1½ hours to accomplish.

Labor units are applied to each item of material and then extended and totaled to give the total worker-hours required to complete the project. The value of the labor in dollars and cents is then determined by multiplying the total computed worker-hours by the installers' average hourly rate of pay.

A separate unit of labor should be provided for the installation of each item of material or labor operation performed. This unit should be broken down further to apply to varying working conditions. For example, if a labor unit is given for installing 100 ft (30 m) of 2-conductor no. 22 cable at ground level, it stands to reason that installing the same amount of cable 20 ft (60 m) above the floor will require more worker-hours. Scaffolding would have to be set up and moved into the area, and the workers would be required to spend more time carrying reels of wire to the scaffold platform. Even if the work could be done with a ladder, some additional time would be required to move the ladder and climb up and down as the movement takes place.

There are several other conditions affecting the labor operation that must be given consideration in preparing any and all bids:

1. The type of building construction
2. Height of the installation above normal working areas
3. The weight of the material or equipment
4. Performance of the general contractor (if any)
5. The availability and proficiency of the workers
6. Whether the wiring is to be concealed or exposed
7. Whether the installation is installed in new or existing buildings.

The basic labor operation for any security/fire-alarm system installation must take into consideration several factors often overlooked by the inexperienced contractor or estimator. For example, the labor unit must include layout instructions, material handling, the actual installation of the material, coffee breaks, visits to the rest room, etc. If the labor units used do not include all of these items, the contract must make allowances to cover them. From the above statements, we can see that the amount of time required for a worker to install a given item may not be an adequate basis for determining an accurate labor unit.

The contractor or estimator is continually faced with having to use good judgment (and educated guesses) when dealing with labor units. At first, the selection of the proper labor units may seem to be a difficult task, but after some experience in the field, the estimator will be able to choose the labor units most applicable to the particular project or portion of a job.

The first step in arriving at the most accurate total estimated labor for a given job is to take off and list the material items on the pricing sheets, segregated in accordance with the installation and building conditions. On larger and more complex jobs, the different categories applicable to each type of material can be expanded in line with the different specific installation conditions.

The second step is to apply the labor unit specifically related to that particular installation condition for the size and type of material involved, depending on the extent of segregated listing of the material and the extent of segregation of the available labor data. However, there is no point in listing the materials on a segregated basis if segregated labor data are not available or if the estimator does not adjust the existing data to account for the specific conditions. Anything less than a segregated listing of the materials in accordance with the varying installation conditions and the application of related labor data reduces the accuracy of the total estimated labor.

Once the choice of labor unit has been made, the mechanics of labor-unit entry consist of merely copying the appropriate labor units from whatever source is available and entering the units in the labor-unit column on the pricing sheet opposite the proper item of material or labor operation.

After all of the labor units have been applied on the pricing sheet, they should be extended and totaled. This operation involves little more than elementary mathematics, but many errors can be made. The estimator, therefore, should be extremely careful at this point. One decimal point in the wrong place can mean the difference between a profit and a loss on a project. It is recommended that a good electronic calculator be used in making all extensions and totals.

No bid should ever be turned in without a check of the figures. Preferably, the person making the initial takeoff should check through the figures; then someone else should quickly check them over. One method of checking column totals is to add them first from top to bottom and then from bottom to top.

In any case, sufficient time should be allowed for checking the figures, as errors often result from hasty last-minute efforts to complete an estimate to meet a specific bid time.

SUMMARIZING THE ESTIMATE

Summarizing the estimate is the final accumulation of all estimated costs such as labor, material, job factors, direct cost, overhead, and profit. Determination of the final quotation is one of the most important steps in preparing the estimate, because one mistake in the final summarizing can affect all of the accuracy with which the previous steps have been handled.

A typical bid summary sheet includes the following basic sections or groups of cost data:

1. Description of the project
2. Cost of listed material and labor
3. Nonproductive labor
4. Direct job expenses
5. Taxes, bonds, etc.
6. Overhead
7. Profit

Such a form serves as a sound guide to accurately summarize the estimate for practically any security/fire-alarm system installation.

DIRECT JOB EXPENSE AND OVERHEAD

A thorough understanding of both direct job expense and overhead is necessary so that they may be included in the final estimate to defray such costs. In general, direct job expenses are those costs (in addition to labor and materials) that have to be paid for as a direct result of performing the job. In other words, if the job was not performed, these costs would not occur. Overhead expense, on the other hand, is all costs that have to be paid whether the particular job is being done or not.

An estimate is not complete until all direct job and overhead expenses have been added to the other items entering into the cost of the project. Direct job expenses are relatively simple to calculate if the contractor is fair with the firm and includes all items of expense that relate directly to the job at hand. Calculating overhead, however, is a different picture altogether. Many contractors take their previous overhead figures and apply them to work which will be performed in the future. This may result in an accurate estimate, but in most cases, the overhead will change during the performance of the work being bid. Therefore, the contractor should analyze the anticipated future overhead for all jobs being bid at the present.

Another consideration is the size of the job. It is a fact that, in most cases, a small job will cause a higher percentage of overhead than a large job. However, the contractor cannot assume that this will always be true, especially in the case of specialized projects.

When the estimate has been completed to the point of adding the overhead, the known data should include cost of materials, cost of labor, and direct job expense. The overhead is then determined by one of the following methods:

1. The overhead for the year may be divided by the gross sales volume for the year to find the overhead as a percentage of the gross sales volume. This percentage is then applied to the prime cost of the job.

2. The overhead expense for the year may be divided by the total cost of labor, material, and job expense for the year to find the overhead as a percentage of the prime cost.

When the estimated annual volume differs from the past annual volume for which an overhead based upon accounting records is obtainable and when, for some reason, the same total dollar cost of overhead expenses must be maintained, the estimator must determine by simple proportion the applicable average overhead percentage of prime cost and apply this percentage using the job-size scale. This is done by estimating the overhead percentage on the basis of past recorded data, adjusted to future volume and size of work. If the job being estimated represents a change in general work pattern or is a special type of job, the estimator must make an intelligent analysis of all the conditions and further adjust the estimated overhead percentage to be applied as accurately as possible.

COMPLETING THE SUMMARY

Completing the summary involves only the inclusion of such miscellaneous items as wire connectors, tape, and fasteners. Most contractors feel that the listing of these items serves no purpose. Therefore, on most projects, an allowance for these items is made rather than a list with the price of each individual item. This allowance is usually in the form of a lump-sum figure, a percentage gained from experience, or an educated guess. As a rule, ½ of 1 percent is sufficient for all projects except highly specialized ones. Once this figure has been determined, the dollar value should be entered in the appropriate space on the summarizing form.

The contractor also will be required to calculate miscellaneous labor costs on many projects. Conditions such as overtime (required to complete the project within a specified time), labor disputes, and special installations will make the inclusion of extra labor necessary. There is no set rule for calculating this figure exactly. It is a matter of experience and good judgment.

The subtotals of the dollar value of the labor, material, subcontractors (if any), and direct job expense are totaled to give the total prime cost. The percentage of applicable estimated overhead determined as previously discussed is applied, and the dollar value of the overhead expense is calculated. This, added to the prime cost, gives the total gross cost.

The percentage of profit to be included in the estimate is either determined by the contractor alone or after consultation between the contractor and the estimator, taking into consideration the type and size of the job, the character of the competition on the job, and the desirability of obtaining the job.

Some contractors prefer to apply a flat percentage of profit to all estimates. Others vary the percentage in accordance with the factors indicated above. Some do not use a percentage adder, but determine the dollar value of the profit desired on the basis of a certain amount for each worker-day required by the job or by allowing a flat sum.

There are certain items of cost that in a true sense are direct job expenses but against which it may not be desirable to assess a profit. Such items may be sales taxes, excise taxes, and payment and performance bonds. If these items have not been included previously in the estimate, they must be added into the final price.

The total estimated price is calculated by totaling the gross cost, profit, and other items. Normally the total estimated price or the nearest even figure is determined to be the amount of the bid. In too many instances, when the contractor or estimator becomes uneasy over the competition on the job, the amount of the bid bears little resemblance to the total estimated price.

Too much emphasis cannot be put on the necessity of including in the summary the proper allowances for direct job expense, job factor, nonproductive labor, labor productivity factor, overhead expenses, and profit. Any estimate properly summarized will more nearly provide enough income from that job to pay for all costs, both direct and indirect, caused by that job than if the final price is established on a hit-or-miss basis. When each job proves to be reasonably profitable, the entire business operation is successful.

LABOR UNITS

The labor units in Fig. 10-1 are the result of averaging the figures of several proven estimating manuals. Bear in mind that these units are based on workers experienced in the security/fire-alarm field and, furthermore, that modern hand-held power tools, wire strippers, and other time-saving devices are assumed to be used in the installation.

Needless to say, labor units are the most important factor in estimating, and no one list of units will accurately forecast the work of all contractors. Worker-hour values are given in Fig. 10-1 for the most commonly used items of security/fire-alarm systems. Interested contractors may adopt these units and modify them appropriately to suit their own operations.

Remember, these labor units reflect the work of experienced contractors. If it

Item to be installed	*Group*			
	1	*2*	*3*	*4*
Alarm bells	0.50	0.70	0.90	1.10
Alarm panels	1.75	2.00	2.25	2.50
Door cords	0.70	1.00	1.30	1.65
Exit/entry delay module	0.45	0.50	0.70	0.85
Fire-alarm stations	0.50	0.70	0.90	1.10
Foil blocks	0.40	0.75	1.00	1.25
Heat sensors	0.80	1.00	1.25	1.50
Horns	0.70	1.00	1.30	1.65
Indoor remote stations	2.50	3.30	5.50	7.70
Indoor sounding devices	.70	1.00	1.30	1.65
Lockswitches	1.00	1.50	2.00	2.50
Low-voltage cable, 2/c (per 1000 ft)	8.50	12.00	14.00	16.00
Low-voltage cable, 3/c (per 1000 ft)	9.50	14.00	16.00	18.00
Low-voltage cable, 4/c (per 1000 ft)	10.50	15.50	17.50	19.50
Magnetic contacts (pair)	0.40	0.75	1.00	1.25
Mechanical contacts	0.30	0.50	0.70	0.90
Motion detectors	1.30	1.80	2.10	2.50
Outdoor alarm bells	1.25	1.75	2.00	2.30
Outdoor remote stations	3.00	3.50	4.00	4.50
Photoelectric cells, recessed	1.20	1.75	2.00	2.30
Photoelectric cells, surface-mounted	1.00	1.50	1.75	2.00
Power supplies	0.30	1.50	2.00	2.50
Relays	1.20	1.70	2.10	2.50
Sirens	1.00	1.50	2.00	2.30
Smoke detectors	0.80	1.00	1.25	1.50
Telephone dialers	2.00	2.50	3.00	3.50
Transformers	0.45	0.50	0.70	0.85
Window foil (per 100 ft)	4.50	5.00	5.50	6.00
Window sensors	0.70	1.00	1.30	1.65

FIGURE 10-1. Security/fire-alarm system labor units (worker-hours).

is necessary to increase any of them drastically, there must be a reason—time-consuming work habits, use of devices or materials that slow down installation, etc. It would be advantageous to seek out these reasons and find ways to improve operating efficiencies. Labor units are given for several groups of difficulty for installations using low-voltage cable, since all degrees of working conditions are likely to be encountered.

There are many installation variables that can affect the labor cost for any given type of outlet, contact installation, wire-pulling operation, etc. A large number of these conditions can be divided into four installation situation groups for all practical purposes. These groups may be designated 1, 2, 3, and 4, with group 1 representing the least amount of work and group 4 requiring the greatest number of worker-hours for the same labor operation.

Group 1: All working areas are open and readily accessible to workers. Work above grade levels requires no scaffolding, only step ladders.

Group 2: Includes the installation of security/fire-alarm equipment, contacts, and wiring in areas that are partially accessible but require minor fishing of cables in concealed partitions. Installation of surface molding to conceal wiring also will fall into this group.

Group 3: These wiring situations usually involve the installation of concealed wiring in partially inaccessible areas, such as crawl spaces, limiting the working room. Other situations include notching of firestops or diagonal bracing to get cables in finished wall spaces, installing wiring on masonry walls where furring strips have been applied, and installing wiring in attics or basements where both horizontal and vertical surfaces have been closed in.

Group 4: The most difficult situations include cutting through masonry walls, removal of finished floor boards to route wiring, removal of baseboards and door/window trim to permit routing of new wiring, or cutting and patching of finished surfaces to conceal new wiring.

PRACTICAL APPLICATION

To illustrate how a typical estimate is performed, take the residence shown in Fig. 10-2. The design criteria are as follows:

1. Security- and fire-alarm system is to be provided.
2. System is to be local type.
3. Distance to street is 50 ft.
4. Distance to neighbors on each side is approximately 150 ft.
5. There is a wooded area in the rear yard.
6. Front and side doors are used for egress and entry.
7. All control stations are to be inside.
8. There are no pets.

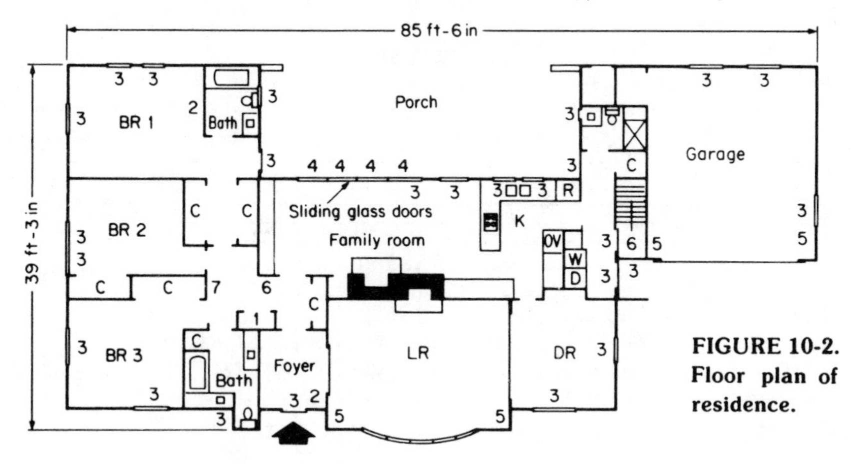

FIGURE 10-2. Floor plan of residence.

9. House is fully carpeted.
10. Security system is to be armed only with windows closed.
11. Rear yard is to be lighted during a security alarm.
12. Window types:
    a. Kitchen: crank-out casement
    b. Living room: bow
    c. All others: double-hung.
13. Door types:
    a. Front: steel with wood casing
    b. All others: wood.

The security/fire-alarm system was laid out as shown in Fig. 10-2. The large numbers correspond to the following equipment:

1. Control panel
2. Remote station (siren speakers in attic at gable louvers)
3. Magnetic contacts
4. Glass-break detectors
5. Photoelectric detector
6. Smoke detector
7. Fire horn.

ESTIMATE SHEET

JOB: Residence, 6111 Heywood Blvd., Kenton, Ky. 60412 — PAGE 1 OF 1 PAGES

ESTIMATED BY: J.E.T. — CHECKED BY: WJN — DATE: 6/7/78

DESCRIPTION	MATERIAL: Quantity	Unit Price	Per	Amount	LABOR (HRS): Unit	Per	Amount
051 Control panel	1	117.60	ea.	117.60	1.75	ea.	1.75
413 Battery	1	12.20		12.70	.30		.30
216R Indoor remote station	3	13.05		39.15	2.50		7.50
290 Prealarm station	1	9.30		9.30	.50		.50
551 Outdoor sounding device	2	19.30		38.60	1.25		2.50
205 Indoor sounding device	1	10.45		10.45	.70		.70
550 Sirendriver	1	20.00		20.00	1.00		1.00
304 Relay	1	5.30		5.30	1.20		1.20
208S Photoelectric system	2	103.50		207.00	1.20		2.40
295A Smoke detector	2	60.12		120.24	.80		1.60
443 Magnetic contacts	25	3.50		87.50	.40		10.00
170 Glassbreak detector	4	15.50		62.00	.70		2.80
121 Adapter (delay)	1	9.80		9.80	1.20		1.20
217 Emergency switch	3	.55		1.65	1.00		3.00
221 Transformer	1	4.00		4.00	.45		.45
18/2 Wire	646 ft	.06	ft.	38.76	8.50	M	5.44
Misc. (fastenings, etc.)	lot			9.24	9.24		4.00
		TOTAL		792.79			46.34

MISCELLANEOUS		RECAPITULATION	
		Material Cost	792.79
		46.34 Hours Labor @ 12.00	556.08
		2.56 NONPRODUCTIVE Hours Labor @ 12.00	30.72
		Direct Job Expense	154.03
		Total Prime Cost	1533.62
		Overhead Expense 15%	230.04
		TOTAL COST	1763.06
		Profit 15%	264.54
		Selling Price	2028.21
Form E-3		BID SUMMITTED	

FIGURE 10-3. Completed estimate sheet.

Begin the material takeoff by listing all major components by catalog number, quantity, and price of each item. Use a rotometer or scale and calculate the total footage of wire. Then use a lump-sum figure for miscellaneous items like connectors, fastenings, etc. When completed, extend the figures as in Fig. 10-3.

Refer to the labor units in Fig. 10-1 and enter proper units in the appropriate column. Extend the labor units, then total both the material and labor columns. Enter the material dollar value and labor worker-hours under the recapitulation columns.

Multiply the worker-hours by the average hourly rate, calculate the labor adder and other direct job expenses (if any), and enter them in the proper spaces. Total the dollar value of material, labor, and direct job expense to obtain a total prime cost. Finish the estimate by adding overhead and profit to obtain a selling price. Figure 10-3 shows that the total selling price is $2,028.21.

COMPUTERIZED ESTIMATING

In the quest to shorten the lengthy process of estimating the installation costs of security fire-alarm systems, two major developments have occurred: the assembly concept and computerized estimating. Both developments have the potential of providing great benefits to installation contractors, although reaping the full benefits of these improvements takes a bit of effort, particularly because it is difficult to tailor the systems to meet the specific needs of each contractor.

The assembly method was first popularized by the Estimatic Corp. in the 1950s. (Estimatic was also a pioneer in computerized estimating long before anyone else was doing it.) The company used the concept that virtually every type of electrical/electronic symbol used in construction drawings could be summarized as a specific list of materials.

For example, the assembly for a common duplex receptacle would include the receptacle, the finish plate, a box, a plaster ring, screws for fastening the box to the framing, a grounding pigtail, a couple of feet of no. 12 wire in the box, an average of two wire nuts, and two ½-in EMT connectors. Thus the assembly includes everything indicated by the duplex receptacle symbol on the plans.

This is typical of all assemblies, where there could be many thousands of combinations of different types of receptacles with different types of finish plates, different types of plaster rings, and so on. Even raceways or cable assemblies can be broken down into assemblies: for example, three no. 12 THHNs in ½-in EMT, three no. 6s and one no. 10 in 1-in EMT, and so on.

Thus the contractor develops a full set of assemblies and prices them all, so that when the estimator takes on a job, the estimator counts all the symbols and raceways on the plans, and then prices the list as assemblies (so many type XYZ assemblies, so many type ABC assemblies, etc.). Because all the assemblies are prepriced and prelabored, the estimator no longer needs to count or price most individual parts. And when it is time to order material, the modern computer can take these assemblies and list how many of each individual item to order from suppliers.

Certainly computerized estimating has made a noticeable impact on the trade, with the computer able to perform all the tedious mathematic functions. It can save an estimator a lot of work and at the same time eliminate the likelihood of errors in mathematics. These are the main benefits of computerized estimating. There are also other benefits to be reaped, but none nearly as important.

The takeoff remains difficult (even with the automatic takeoff tools). Then comes writing up the estimate; that is, entering it into the computer. At first glance this would seem far easier than the "old" method of writing down each item and assigning it a cost and labor rate, but other factors come into play with the computer that diminish the gain somewhat.

First of all there are the computer codes. All the popular computerized estimating systems use a special code number for each particular type and size of material. For example, ½-in EMT might be code number 27.7; 1-in EMT might be code number 27.11; no. 12 THHN solid could be code number 246.94, and so on. So now the estimator must enter not only a description of the item, but a code number also. (This is not always necessary, but almost all estimators enter a description so they can double-check their work.) And because there are so many different items in electrical construction that no one remembers all the code numbers, it is necessary for the estimator to look up code numbers for most of the items. Some estimating rooms have organized a large chart on the walls from which the estimator can obtain all the code numbers at a glance.

Thus far the best method developed for itemized estimating is to use a combination of assemblies and the computer. In the best of these systems the estimator does the standard type takeoff and enters the totals into the computer. Then the computer takes all these assemblies, breaks them down into their individual parts, and prints out one complete bill of material.

You get the best of the assembly system's benefits, and you don't have to update the prices of your assemblies continually, because all the computer needs is the quantities of materials that make up each assembly, not the prices. After the whole job is broken down to a single material list, the computer will give a material price and labor hour figure to each item, and not to entire assemblies.

Of course, the task of updating the material prices in the computer remains; but given the makeup of the industry at this time, there is no worthwhile alternative to spending some time doing this.

Although computerized estimating systems have been around for a number of years, only in the past decade have the prices of both the computer equipment and the programs decreased to where the average contractor can afford them. Actually, computers for the past few years have been making the transition from a "nice tool" to "standard equipment" in electrical contractors' offices.

Remember when purchasing any computer system (but especially for estimating) that flexibility is critically important, as is proper training in the use of the system. Any computer system that you use for estimating should be capable of being fully integrated with computer programs your company may be using for accounting and management.

GLOSSARY

Accessible (as applied to equipment) Admitting close approach because not guarded by locked doors, elevation, or other effective means.

Accessible (as applied to wiring methods) Capable of being removed or exposed without damaging the building structure or finish, or not permanently closed in by the structure or finish of the building.

AC *See* Alternating current.

Accelerator (1) A substance that increases the speed of a chemical reaction. (2) Something to increase velocity.

Alternating current (ac) (1) A periodic current, the average of which is zero over a period; normally the current reverses after given time intervals and has alternately positive and negative values. (2) The type of electrical current actually produced in a rotating generator (alternator).

Alternator A device to produce alternating current.

Ambient temperature Temperature of fluid (usually air) that surrounds an object on all sides.

American National Standards Institute (ANSI) An organization that publishes nationally recognized standards.

American Wire Gage (AWG) The standard for measuring wires in America.

Ammeter An electric meter used to measure current; it is calibrated in amperes.

Ampacity The current-carrying capacity of conductors or equipment, expressed in amperes.

Ampere (A) The basic SI unit measuring the quantity of electricity.

Ampere-turn The product of amperes times the number of turns in a coil.

Amplification Procedure of expanding the strength of a signal.

Amplifier (1) A device that enables an input signal to directly control a larger energy flow. (2) The process of increasing the strength of an input.

Amplitude The maximum value of a wave.

Analog Pertaining to data from continuously varying physical quantities.

ANSI *See* American National Standards Institute.

Antenna A device for transmission or reception of electromagnetic waves.

Appliance Equipment designed for a particular purpose that uses electricity to produce heat, light, mechanical motion, etc., usually complete in itself, generally other than industrial, normally in standard sizes or types.

Approved (1) Acceptable to the authority having legal enforcement. (2) A product approved by OSHA; one that has been tested to standards and found suitable for general application, subject to limitations outlined in the nationally recognized testing laboratory's listing.

Arc A flow of current across an insulating medium.

Arcing time The time elapsing from the severance of the circuit to the final interruption of current flow.

Arc resistance The time required for an arc to establish a conductive path in or across a material.

Armature (1) Rotating machine: the member in which alternating voltage is generated. (2) Electromagnet: the member that is moved by magnetic force.

Armor Mechanical protector for cables; usually a helical winding of metal tape, formed so that each convolution locks mechanically on the previous one (interlocked armor); may be a formed metal tube or a helical wrap of wires.

Arrester Wire screen secured to the top of an incinerator to confine sparks and other products of burning.

Attachment plug or cap The male connector for electrical cords.

Attenuation A decrease in energy magnitude during transmission.

Audible Capable of being heard by humans.

Automatic Operating by own mechanism when actuated by some impersonal influence: nonmanual: self-acting.

Automatic transfer equipment A device to transfer a load from one power source to another, usually from normal to emergency source and back.

Autotransformer Any transformer where primary and secondary connections are made to a single cell.

Auxiliary A device or equipment which aids the main device or equipment.

AWG *See* American Wire Gage.

Ballast A device designed to stabilize current flow.

Battery A device which changes chemical to electrical energy, used to store electricity.

Bimetal strip Temperature regulating or indicating device that works on the principle that two dissimilar metals with unequal expansion rates, welded together, will bend as temperature changes.

Bonding bushing A special conduit bushing equipped with a conductor terminal to take a bonding jumper; also has a screw or other sharp device to bite into the enclosure wall to bond the conduit to the enclosure without a jumper when there are no concentric knockouts left in the wall of the enclosure.

Bonding jumper A bare or insulated conductor used to ensure the required electrical conductivity between metal parts required to be electrically connected. Frequently used from a bonding bushing to the service equipment enclosure to provide a path around concentric knockouts in an enclosure wall; also used to bond one raceway to another.

Bonding locknut A threaded locknut for use on the end of a conduit terminal, but a locknut equipped with a screw through its lip. When the locknut is installed, the screw is tightened so its end bites into the wall of the enclosure close to the edge of the knockout.

Braid An interwoven cylindrical covering of fiber or wire.

Branch circuit That portion of a wiring system extending beyond the final overcurrent device protecting a circuit.

Bridge A circuit which measures by balancing four impedances through which the same current flows:

Wheatstone—resistance
Kelvin—low resistance
Schering—capacitance, dissipation factor, dielectric constant
Wien—capacitance, dissipation factor

Bus The conductor(s) serving as a common connection for two or more circuits.

Bus bars The conductive bars used as the main current supplying elements of panelboards or switchboards; also the conductive bars duct: an assembly of bus bars within an enclosure which is designed for ease of installation, has no fixed electrical characteristics, and allows power to be taken off conveniently, usually without circuit interruption.

BX A nickname for armored cable (wires with a spiral-wound, flexible steel outer jacketing); although used generically, **BX** is a registered tradename of the General Electric Company.

Bypass Passage at one side of or around a regular passage.

Cable An assembly of two or more wires which may be insulated or bare.

Cable, aerial An assembly of one or more conductors and a supporting messenger.

Cable, armored A cable having armor. (*See* Armor.)

Cable, belted A multiconductor cable having a layer of insulation over the assembled insulated conductors.

Cable clamp A device used to clamp around a cable to transmit mechanical strain to all elements of the cable.

Cable, coaxial A cable used for high frequency, consisting of two cylindrical conductors with a common axis separated by a dielectric; normally the outer conductor is operated at ground potential for shielding.

Cable, control Used to supply voltage (usually ON or OFF).

Cable, duplex A twisted pair of cables.

Cable, power Used to supply current (power).

Cable, pressure A cable having a pressurized fluid (gas or oil) as part of the insulation; paper and oil are the most common insulators.

Cable, ribbon A flat multiconductor cable.

Cable, service drop The cable from the utility line to the customer's property.

Cable, signal A cable used to transmit data.

Cable, spacer An aerial distribution cable made of covered conductors held by insulated spacers; designed for wooded areas.

Cable tray A rigid structure to support cables: a type of raceway: normally having the appearance of a ladder and open at the top to facilitate changes.

Cable, tray A multiconductor having a nonmetallic jacket, designed for use in cable trays (not to be confused with type TC cable, for which the jacket must also be flame retardant).

Cable, triplexed d Helical assembly of three insulated conductors and sometimes a bare grounding conductor.

Cable, unit A cable having pairs of cables stranded into groups (units) of a given quantity; these groups then form the core.

Cable, vertical riser Cables utilized in circuits of considerable elevation change; usually incorporate additional components for tensile strength.

Cabling Helically wrapping together of two or more insulated wires.

Capacitance The storage of electricity in a capacitor; the opposition to voltage change; the unit of measurement is the farad.

Capacitor An apparatus consisting of two conducting surfaces separated by an insulating material. It stores energy, blocks the flow of direct current, and permits the flow of alternating current to a degree depending on the capacitance and frequency.

Capillary action The traveling of liquid along a small interstice due to surface tension.

Capstan A rotating drum used to pull cables or ropes by friction; the cables are wrapped around the drum.

Cathode (1) The negative electrode through which current leaves a nonmetallic conductor, such as an electrolytic cell. (2) The positive pole of a storage battery. (3) Vacuum tube—the electrode that emits electrons.

Cathode-ray tube The electronic tube which has a screen upon which a beam of electrons from the cathode can be made to create images; for example, the television picture tube.

Cathodic protection Reduction or prevention of corrosion by making the metal to be protected the cathode in a direct current circuit.

Cavity wall Wall built of solid masonry units arranged to provide airspace within the wall.

CB Pronounced "see bee," an expression used to refer to "circuit breaker," taken from the initial letters C and B.

C-C Center to center.

CCA *See* Customer Complaint Analysis.

CEE *See* International Commission on Rules for the Approval of Electrical Equipment.

Centigrade scale Temperature scale used in metric system. Freezing point of water is 0°; boiling point is 100°.

CFR *See* Code of Federal Regulations.

Choke coil A coil used to limit the flow of alternating current while permitting direct current to pass.

Circuit A closed path through which current flows from a generator, through various components, and back to the generator.

Circuit breaker A resettable fuse-like device designed to protect a circuit against overloading.

Circuit foot One foot of circuit; i.e., if one has a three-conductor circuit, each lineal foot of circuit would have three circuit feet.

Circular mil The non-SI unit for measuring the cross-sectional area of a conductor.

CL Center line.

Clearance The vertical space between a cable and its conduit.

Coaxial cable A cable consisting of two conductors concentric with and insulated from each other.

Code Short for National Electrical Code.

Code installation An installation that conforms to the local code and/or the national code for safe and efficient installation.

Code of Federal Regulations (CFR) The general and permanent rules published in the Federal Register by the executive departments and agencies of the federal government. The Code is divided into 50 titles, which represent broad areas; titles are divided into chapters, which usually bear the name of the issuing agency; e.g., Title 30—Mineral Resources, Chapter 1—MESA; Title 29—Labor, Chapter XVII—OSHA; Title 10—Energy, Chapter I—NRC.

Color code Identifying conductors by the use of color.

Come along A cable grip (usually of tubular basket-weave construction which tightens its grip on the cable as it is pulled) with a pulling "eye" on one end for attaching to a pull-rope for pulling conductors into a conduit or other raceway.

Computer An electronic apparatus: (1) for solving complex and involved problems, usually mathematical or logical, rapidly, (2) for storing large amounts of data.

Concealed Rendered inaccessible by the structure or finish of the building. Wires in concealed raceways are considered concealed, even though they may become accessible by being withdrawn.

Concentricity The measurement of the center of the conductor with respect to the center of the insulation.

Conductance The ability of material to carry an electric current.

Conductor Any substance that allows energy flow through it, with the transfer being made by physical contact but excluding net mass flow.

Conductor, bare Having no covering or insulation whatsoever.

Conductor, covered A conductor having one or more layers of nonconducting materials that are not recognized as insulation under the National Electrical Code.

Conductor, insulated A conductor covered with material recognized as insulation.

Conductor load The mechanical load on an aerial conductor—wind, weight, ice, etc.

Conductor, plain A conductor that consists of only one metal.

Conductor, segmental A conductor that has sections isolated one from the other and connected in parallel; used to reduce ac resistance.

Conductor, solid A single wire.

Conductor, stranded Assembly of several wires, usually twisted or braided.

Conductor stress control The conducting layer applied to make the conductor a smooth surface in intimate contact with the insulation; formerly called extruded strand shield (ESS).

Conduit A tubular raceway.

Conduit fill Amount of cross-sectional area used in a raceway.

Conduit, rigid metal Conduit made of Schedule 40 pipe, normally 10-ft lengths.

Configuration, cradled The geometric pattern which cables will take in a conduit when the cables are pulled in parallel and the ratio of the conduit ID to the 1/C cable OD is greater than 3.0.

Configuration, triangular The geometric pattern which cables will take in a conduit when the cables are triplexed or are pulled in parallel with the ratio of the conduit ID to the 1/C cable OD less than 2.5.

Connection (1) The part of a circuit which has negligible impedance and which joins components or devices; (2) A cable terminal, splice, or seal at the interface of the cable and equipment.

Connection, delta Interconnection of three electrical equipment windings in delta (triangular) fashion.

Connection, star Interconnection of three electrical equipment windings in star (wye) fashion.

Connector A device used to physically and electrically connect two or more conductors.

Connector, pressure A connector applied by using pressure to form a cold weld between the conductor and the connector.

Connector, reducing A connector used to join two different size conductors.

Constant current A type of power system in which the same amount of current flows through each utilization equipment, used for simplicity in street-lighting circuits.

Constant voltage The common type of power in which all loads are connected in parallel, but different amounts of current flow through each load.

Contact A device designed for repetitive connections.

Contactor A type of relay.

Continuity The state of being whole, unbroken.

Continuous load (1) As stipulated by NEC—in operation 3 hours or more. (2) For nuclear power—8760 hours/year (scheduled maintenance outages permitted).

Continuous vulcanization (CV) A system utilizing heat and pressure to vulcanize insulation after extrusion onto wire or cable; the curing tube may be in a horizontal or a vertical pole.

Control Automatic or manual device used to stop, start, and/or regulate flow of gas, liquid, and/or electricity.

Copper A word used by itself to refer to copper conductors. Examples: "A circuit of 500 MCM copper" or "the copper cost of the circuit." It is a good conductor of electricity, easily formed, and easily connected to itself and other metals.

Cord A small flexible conductor assembly, usually jacketed.

Cord set A cord having a wiring connector on one or more ends.

Core The portion of a foundry mold that shapes the interior of a hollow casting.

Core (cable) The portion of an insulated cable under a protective covering.

Counter emf The voltage opposing the applied voltage and the current in a coil; caused by a flow of current in the coil; also known as back emf.

Coupling The means by which signals are transferred from one circuit to another.

Coupon A piece of metal for testing, of specified size; a piece of metal from which a test specimen may be prepared.

CT Pronounced "see tee," refers to current transformer, taken from the initial letters C and T.

CU Copper.

Current (I) The time rate of flow of electric charges; measured in amperes.

Current, charging The current needed to bring the cable up to voltage; determined by capacitance of the cable; after withdrawal of voltage, the charging current returns to the circuit; the charging current will be 90° out of phase with the voltage.

Current density The current per unit cross-sectional area.

Current-induced Current in a conductor due to the application of a time-varying electromagnetic field.

Current, leakage The small amount of current which flows through insulation whenever a voltage is present and heats the insulation because of the insulation's resistance; the leakage current is in phase with the voltage, and is a power loss.

Current limiting A characteristic of short-circuit protective devices, such as fuses, by which the device operates so fast on high short-circuit currents that less than a quarter wave of the alternating cycle is permitted to flow before the circuit is opened, thereby limiting the thermal and magnetic energy to a certain maximum value, regardless of the current available.

Customer Complaint Analysis (CCA) A formal investigation of a cable defect or failure.

Cut in The connection of electrical service to a building, from the power company line to the service equipment, e.g., "the building was cut in" or "the power company cut in the service."

Cycle (1) An interval of space or time in which one set of events or phenomena is completed. (2) A set of operations that are repeated regularly in the same sequence. (3) A number of different processes a system in a given state goes through before it finally returns to its initial state.

Dead (1) Not having electrical charge. (2) Not having voltage applied.

Dead-end A mechanical terminating device on a building or pole to provide support at the end of an overhead electric circuit. A dead-end is also the term used to refer to the last pole in the pole line. The pole at which the electric circuiting is brought down the pole to go underground or to the building served.

Dead-front A switchboard or panel or other electrical apparatus without "live" energized terminals or parts exposed on the front, where personnel might make contact.

Demand (1) The measure of the maximum load of a utility's customer over a short period of time. (2) The load integrated over a specified time interval.

Demand factor For an electrical system or feeder circuit, a ratio of the amount of connected load (in kVA or amperes) which will be operating at the same time to the total amount of connected load on the circuit. An 80% demand factor, for instance, indicates that only 80% of the connected load on a circuit will ever be operating at the same time. Conductor capacity can be based on that amount of load.

Detection The process of separating the modulation component from the received signal.

Device An item intended to carry, or help carry, but not utilize electrical energy.

Dew point The temperature at which vapor starts to condense (liquify) from a gas-vapor mixture at constant pressure.

Dielectric strength The maximum voltage which an insulation can withstand without breaking down; usually expressed as a gradient, in volts per mil (vpm).

Diode A device having two electrodes, the cathode and the plate or anode, and which is used as a rectifier and detector.

Direct current (dc) (1) Electricity which flows in only one direction. (2) The type of electricity produced by a battery.

Disconnect A switch for disconnecting an electrical circuit or load (motor, transformer, panel) from the conductors which supply power to it; e.g., "He pulled the motor disconnect" means he opened the disconnect switch to the motor.

Disconnecting means A device, a group of devices, or other means whereby the conductors of a circuit can be disconnected from their supply source.

Distribution, statistical analysis A statistical method used to analyze data by correlating data to a theoretical curve in order to (a) test validity of data; (b) predict performance at conditions different from those used to produce the data. The normal distribution curve is most common.

Drawing, block diagram A simplified drawing of a system showing major items as blocks; normally used to show how the system works and what the relationship between major items is.

Drawing, line schematic (diagram) Shows how a circuit works.

Drawing, plot or layout Shows the "floor plan."

Drawing, wiring diagram Shows how the devices are interconnected.

Drill A circular tool used for machining a hole.

Drywall Interior wall construction consisting of plasterboard, wood paneling, or plywood nailed directly to the studs without application of plaster.

Duty, continuous A service requirement that demands operation at a substantially constant load for an indefinitely long time.

Duty, intermittent A service requirement that demands operation for alternate intervals of load and no load, load and rest, or load, no load, and rest.

Duty, periodic A type of intermittent duty in which the load conditions regularly reoccur.

Duty, short-time A requirement of service that demands operations at loads and for intervals of time that may both be subject to wide variation.

Edison base The standard screw base used for ordinary lamps.

EEI Edison Electric Institute.

Efficiency The ratio of the output to the input.

Elasticity The property of recovery to original size and shape after deformation.

Electrolyte A liquid or solid that conducts electricity by the flow of ions.

Electrolytic condenser-capacitor Plate or surface capable of storing small electrical charges. Common electrolytic condensers are formed by rolling thin sheets of foil between insulating materials. Condenser capacity is expressed in microfarads.

Electromagnet A device consisting of a ferromagnetic core and a coil that produces appreciable magnetic effects only when an electric current exists in the coil.

Electromotive force (emf) voltage Electrical force that causes current (free electrons) to flow or move in an electrical circuit. The unit of measurement is the volt.

Electron The subatomic particle that carries the unit negative charge of electricity.

Electron emission The release of electrons from the surface of a material into surrounding space due to heat, light, high voltage, or other causes.

Electronics The science dealing with the development and application of devices and systems involving the flow of electrons in vacuum, gaseous media, and semiconductors.

Emitter The part of a transistor that emits electrons.

Engine An apparatus which converts heat to mechanical energy.

Environment (1) The universe within which a system must operate. (2) All the elements over which the designer has no control and that affect a system or its inputs and outputs.

Equipment A general term including material, fittings, devices, appliances, fixtures, apparatus, and the like used as part of, or in connection with, an electrical installation.

Farad The basic unit of capacitance: one farad equals one coulomb per volt.

Fatigue The weakening or breakdown of a material due to cyclic stress.

Fault An abnormal connection in a circuit.

Fault, arcing A fault having high impedance causing arcing.

Fault, bolting A fault of very low impedance.

Fault, ground A fault to ground.

Feedback The process of transferring energy from the output circuit of a device back to its input.

Feeder A circuit, such as conductors in conduit or a busway run, which carries a large block of power from the service equipment to a subfeeder panel or a branch circuit panel or to some point at which the block or power is broken down into smaller circuits.

Fish tape A flexible metal tape for fishing through conduits or other raceway to pull in wires or cables; also made in nonmetallic form of "rigid rope" for hand fishing of raceways.

Fitting An accessory such as a locknut, bushing, or other part of a wiring system that is intended primarily to perform a mechanical rather than an electrical function.

Flex Common term used to refer to flexible metallic conduit.

Flexural strength The strength of a material in bending, expressed as the tensile stress of the outermost fibers of a bent test sample at the instant of failure.

Frequency The number of complete cycles an alternating electric current, sound wave, or vibrating object undergoes per second.

Friction tape An insulating tape made of asphalt-impregnated cloth; used on 600V cables.

Fuse A protecting device which opens a circuit when the fusible element is severed by heating due to overcurrent passing through. Rating: voltage, normal current, maximum let-through current, time delay of interruption.

Fuse, dual element A fuse having two fuse characteristics; the usual combination is having an overcurrent limit and a time delay before activation.

Fuse, nonrenewable or one-time A fuse which must be replaced after it interrupts a circuit.

Fuse, renewable link A fuse which may be reused after current interruption by replacing the meltable link.

Fusible plug A plug or fitting made with a metal of a known low melting temperature; used as a safety device to release pressures in case of fire.

Galvanometer An instrument for indicating or measuring a small electrical current by means of a mechanical motion derived from electromagnetic or dynamic forces.

Gauge (1) Dimension expressed in terms of a system of arbitrary reference numbers; dimensions expressed in decimals are preferred. (2) To measure.

Generator (1) A rotating machine to convert from mechanical to electrical energy. (2) A machine to convert automotive-mechanical to direct current. (3) General apparatus, equipment, etc., to convert or change energy from one form to another.

GFI *See* Ground Fault Interrupter.

Greenfield Another name for flexible metal conduit.

Grommet A plastic, metal, or rubber doughnut-shaped protector for wires or tubing as they pass through a hole in an object.

Ground A large conducting body (as the earth) used as a common return for an electric circuit and as an arbitrary zero of potential.

Ground check A pilot wire in portable cables to monitor the grounding circuit.

Ground coil A heat exchanger buried in the ground that may be used either as an evaporator or a condenser.

Grounded Connected to earth.

Grounded conductor A system or circuit conductor that is intentionally grounded.

Ground Fault Interrupter (GFI) A protective device that detects abnormal current flowing to ground and then interrupts the circuit.

Grounding The device or conductor connected to ground designed to conduct only under abnormal conditions.

Grounding conductor A conductor used to connect metal equipment enclosures and/or the system grounded conductor to a grounding electrode, such as the ground wire run to the water pipe at a service; also may be a bare or insulated conductor used to ground motor frames, panel boxes, and other metal equipment enclosures used throughout an electrical system. In most conduit systems, the conduit is used as the ground conductor.

Grounds Narrow strips of wood nailed to walls as guides to plastering and as a nailing base for interior trim.

Guard A conductor situated so as to conduct interference to its source and prevent the interference from having an influence on the desired signal. (2) A mechanical barrier against physical contact.

Half effect The changing of current density in a conductor due to a magnetic field extraneous to the conductor.

Half wave Rectifying only half of a sinusoidal ac supply.

Handy box The single-gang outlet box which is used for surface mounting to enclose wall switches or receptacles, on concrete or cinder block construction of industrial and commercial buildings; also made for recessed mounting; also known as a utility box.

Hard drawn A relative measure of temper; drawn to obtain maximum strength.

Hardness Resistance to plastic deformation; resistance to scratching, abrasion, or cutting.

Harmonic An oscillation whose frequency is an integral multiple of the fundamental frequency.

Harness A group of conductors laced or bundled in a given configuration, usually with many breakouts.

Heat dissipation The flow of heat from a hot body to a cooler body by (1) convection, (2) radiation, or (3) conduction.

Helix The path followed when winding a wire or strip around a tube at a constant angle.

Henry The derived SI unit for inductance: one henry equals one weber per ampere.

Home run The part of a branch circuit that goes from the panelboard housing the branch circuit fuse or CB and the first junction box at which the branch circuit is spliced to lighting or receptacle devices or to conductors which continue the branch circuit to the next outlet or junction box. The term "home run" is usually reserved for multioutlet lighting and appliance circuits.

Horsepower The non-SI unit for power: 1 hp = 746 W (electric) = 9800 W (boiler).

Hot Energized with electricity.

Hot junction The part of the thermoelectric circuit which releases heat.

Hot leg A circuit conductor which normally operates at a voltage above ground; the phase wires or energized circuit wires other than a grounded neutral wire or grounded phase leg.

IBEW International Brotherhood of Electrical Workers.

IC Pronounced "eye see": Refers to interrupting capacity of any device required to break current (switch, circuit breaker, fuse, etc.), taken from the initial letters *I* and *C*; it is the amount of current the device can interrupt without damage to itself.

ID Inside diameter.

Identified Marked to be recognized as grounded.

IEC International Electrochemical Commission.

IEEE Institute of Electrical and Electronics Engineers.

Ignition transformer A transformer designed to provide a high voltage current.

Impedance (A) The opposition to current flow in an ac circuit; impedance includes resistance (R), capacitive reactance (xc), and inductive reactance (XL); it is measured in ohms.

Impedance matching Matching source and load impedance for optimum energy transfer with minimum distortion.

Impulse A surge of unidirectional polarity.

Inductance The creation of a voltage from a time-varying current; the opposition to current change, causing current changes to lag behind voltage changes; the unit of measurement is a henry.

Infrared lamp An electrical device that emits infrared rays, which are invisible rays just beyond red in the visible spectrum.

Infrared radiation Radiant energy given off by heated bodies which transmits heat and will pass through glass.

In phase The condition existing when waves pass through their maximum and minimum values of like polarity at the same instant.

Instrument A device for measuring the value of the quantity under observation.

Insulated Separated from other conducting surfaces by a substance permanently offering a high resistance to the passage of energy through the substance.

Insulated Power Cable Engineers Association (IPCEA) The association of cable manufacturing engineers who make nationally recognized specifications and tests for cables.

Insulation, class rating A temperature rating descriptive of classes of insulations for which various tests are made to distinguish the materials; not necessarily related to operating temperatures.

Insulation dc resistance constant (IRK) A system to classify materials according to their resistance on a 1000-ft basis at 15.5°C (60°F).

Insulation, electrical A medium in which it is possible to maintain an electrical field with little supply of energy from additional sources; the energy required to produce the electric field is fully recoverable only in a complete vacuum (the ideal dielectric) when the field or applied voltage is removed: used to (a) save space, (b) enhance safety, (c) improve appearance.

Insulation fall-in The filling of strand interstices, especially the inner interstices, which may contribute to connection failures.

Insulation level (cable) The thickness of insulation for circuits having ground fault detectors which interrupt fault currents within (1) 1 minute = 100% level, (2) 1 hour = 133% level, (3) more than 1 hour = 173% level.

Insulation resistance (IR) The measurement of the dc resistance of insulating material; can be either volume or surface resistivity; extremely temperature sensitive.

Insulation, thermal Substance used to retard or slow the flow of heat through a wall or partition.

Integrated circuit A circuit in which different types of devices such as resistors, capacitors, and transistors are made from a single piece of material and then connected to form a circuit.

Integrator Any device producing an output proportionate to the integral of one variable with respect to a second variable; the second variable is usually time.

Intercalated tapes Two or more tapes of different materials helically wound and overlapping on a cable to separate the materials.

Interconnected system Operating with two or more power systems connected through tie lines.

Interference Extraneous signals which are undesired.

Interlock A safety device to ensure that a piece of apparatus will not operate until certain conditions have been satisfied.

International Commission on Rules for the Approval of Electrical Equipment (CEE) Controls the standards for electrical products for sale in Europe; analogous to UL in USA.

Inverter An item which changes dc to ac.

Ion An electrically charged atom or radical.

Ionization (1) The process or the result of any process by which a neutral atom or molecule acquires charge. (2) A breakdown that occurs in gaseous parts of an insulation when the dielectric stress exceeds a critical value without initiating a complete breakdown of the insulation system; ionization is harmful to living tissue, and is detectable and measurable; may be evidenced by corona.

Ionization factor The difference between the percentages of dissipation factors at two specified values of electrical stress; the lower of the two stresses is usually so selected that the effect of the ionization on the dissipation factor at this stress is negligible.

IPCEA *See* Insulated Power Cable Engineers Association.

IR *See* Insulation resistance.

IR drop The voltage drop across a resistance due to the flow of current through the resistor.

IRK *See* Insulation dc resistance constant.

Isolated Not readily accessible to persons unless special means of access are used.

Isolating With switches, means that the switch is not a loadbreak type and must be opened only when no current is flowing in the circuit. This term also refers to transformers (an isolating transformer) used to provide magnetic isolation of one circuit from another, thereby breaking a metallic conductive path between the circuits.

Jacket A nonmetallic polymeric close-fitting protective covering over cable insulation; the cable may have one or more conductors.

Jacket, conducting An electrically conducting polymeric covering over an insulation.

Jumper A short length of conductor, usually a temporary connection.

Junction A connection of two or more conductors.

Junction box Group of electrical terminals housed in a protective box or container.

Kilowatt Unit of electrical power equal to 1000 watts.

Kilowatt-foot The product of load in kilowatts and the circuit's distance over which a load is carried in feet; used to compute voltage drop.

Kinetic energy Energy by virtue of motion.

Kirchhoff's Laws (1) The algebraic sum of the currents at any point in a circuit is zero. (2) The algebraic sum of the product of the current and the impedance in each conductor in a circuit is equal to the electromotive force in the circuit.

Knockout A portion of an enclosure designed to be readily removed for installation of a raceway.

KO Pronounced "kay oh," a knockout, the partially cut opening in boxes, panel cabinets, and other enclosures that can easily be knocked out with a screwdriver and hammer to provide a clean hole for connecting conduit, cable, or some fittings.

KVA Kilovolts times ampere.

LA Lightning arrestor.

Labeled Items carrying the trademark of a nationally recognized testing laboratory.

Leakage Undesirable conduction of current.

Leakage distance The shortest distance along an insulation surface between conductors.

Leg A portion of a circuit.

Lighting outlet An outlet intended for the direct connection of a lamp holder, lighting fixture, or pendant cord terminating in a lamp holder.

Lightning arrestor A device designed to protect circuits and apparatus from high transient voltage by diverting the overvoltage to ground.

Limit control Control used to open or close electrical circuits as temperature or pressure limits are reached.

Limiter A device in which some characteristic of the output is automatically prevented from exceeding a predetermined value.

Line (1) A circuit between two points. (2) Ropes used during overhead construction.

Live-front Any panel or other switching and protection assembly, such as a switchboard or motor control center, which has exposed electrically energized parts on its front, presenting the possibility of contact by personnel.

Live load Any load on a structure other than a dead load; includes the weight of persons occupying the building and freestanding material.

Load (1) A device that receives power. (2) The power delivered to such a device.

Load center An assembly of circuit breakers or switches.

Load factor The ratio of the average to the peak load over a period.

Load losses Those losses incidental to providing power.

Lug A device for terminating a conductor to facilitate the mechanical connection.

Magnet A body that produces a magnetic field external to itself; magnets attract iron particles.

Magnetic field (1) A magnetic field is said to exist at the point at which a force over and above any electrostatic force is exerted on a moving charge at that point. (2) The force field established by ac through a conductor, especially a coiled conductor.

Magnetic pole Those portions of the magnet toward which the external magnetic induction appears to converge (south) or diverge (north).

MCM An expression referring to conductors of sizes from 250 MCM, which stands for thousand circular mils, up to 2000 MCM.

Medium hard A relative measure of conductor temper.

Megger The term used to identify a test instrument for measuring the insulation resistance of conductors and other electrical equipment; specifically, a megohm (million ohms) meter; but Megger is a registered trade name of the James Biddle Co.

Megohmmeter An instrument for measuring extremely high resistance.

Metal clad (MC) The cable core is enclosed in a flexible metal covering.

Mica A silicate which separates into layers and has high insulation resistance, dielectric strength, and heat resistance.

MI cable Mineral-insulated, metal-sheathed cable.

Microwave Radio waves of frequencies above one gigahertz.

Mil A unit used in measuring the diameter of wire, equal to 0.001 in (25.4 μm).

MIL Military specification.

Millimeter (mm) One-thousandth of a meter.

Mil scale The heavy oxide layer formed during hot fabrication or heat treatment of metals.

Modem Equipment that connects data transmitting/receiving equipment to telephone lines: a word contraction of modulator-demodulator.

Modulation The varying of a "carrier" wave characteristic by a characteristic of a second "modulating" wave.

Moisture-resistance So constructed or treated that moisture will not readily injure.

Molded case breaker A circuit breaker enclosed in an insulating housing.

Motor An apparatus to convert from electrical to mechanical energy.

Motor, capacitor A single-phase induction motor with an auxiliary starting winding connected in series with a condenser for better starting characteristics.

Motor control Device to start and/or stop a motor at certain temperature or pressure conditions.

Mutual inductance The condition of voltage in a second conductor because of a change in current in another, adjacent conductor.

National Electrical Code (NEC) A national consensus standard for the installation of electrical systems.

National Fire Protection Association (NFPA) An organization to promote the science and improve the methods of fire protection; it sponsors various codes, including the National Code.

Natural convection Movement of a fluid or air caused by temperature change.

Negative Connected to the negative terminal of a power supply.

NEMA National Electrical Manufacturers Association.

Neoprene An oil-resistant synthetic rubber used for jackets; originally a DuPont trade name, now a generic term for polychloroprene.

Neutral The element of a circuit from which other voltages are referenced with respect to magnitude and time displacement in steady-state conditions.

Neutral block The neutral terminal block in a panelboard, meter enclosure, gutter, or other enclosure in which circuit conductors are terminated or subdivided.

Neutral wire A circuit conductor which is common to the other conductors of the circuit, having the same voltage between it and each of the other circuit wires and usually operating grounded; such as the neutral of three-wire, single-phase, or three-phase, four-wire wye systems.

NFPA *See* National Fire Protection Association.

Nineteen hundred box A commonly used term to refer to any two-gang 4-in square outlet box used for two wiring devices or for one wiring device with a single-gang cover where the number of wires requires this box capacity.

Nipple A threaded pipe or conduit less than 2 ft long.

Occupational Safety and Health Act (OSHA) Federal Law #91-596 of 1970 charging all employers engaged in business affecting interstate commerce to be responsible for providing a safe working place: it is administered by the Department of Labor. The OSHA regulations are published in Title 29, Chapter XVII, Part 1910 of the CFR and the Federal Register.

Ohmmeter An instrument for measuring resistance in ohms.

Ohm's Law Mathematical relationship between voltage, current, and resistance in an electric circuit.

Oscillation The variation, usually with time, of the magnitude of a quantity which is alternately greater and smaller than a reference.

Oscillator A device that produces an alternating or pulsating current or voltage electronically.

Oscillograph An instrument primarily for producing a graph of rapidly varying electrical quantities.

Oscilloscope An instrument primarily for making visible rapidly varying electrical quantities: oscilloscopes function similarly to TV sets.

OSHA *See* Occupational Safety and Health Act.

Outlet A point on the wiring system at which current is taken to supply utilization equipment.

Outline lighting An arrangement of incandescent lamps or gaseous tubes to outline and call attention to certain features, such as the shape of a building or the decoration of a window.

Output (1) The energy delivered by a circuit or device. (2) The terminals for such delivery.

Overload Load greater than the load for which the system or mechanism was intended.

Overvoltage (cable) Voltage above normal operating voltage, usually due to: (a) switching loads on/off, (b) lighting, (c) single phasing.

Pad-mounted A shortened expression for "pad-mount transformer," which is a completely enclosed transformer mounted outdoors on a concrete pad, without need for a surrounding chain-link fence around the metal, box-like transformer enclosure.

Panelboard A single panel or group of panel units designed for assembly in the form of a single panel; includes buses and may come with or without switches and/or automatic overcurrent protective devices for the control of light, heat, or power circuits of individual as well as aggregate capacity. It is designed to be placed in a cabinet or cutout box that is in or against a wall or partition and is accessible only from the front.

Phase conductor Any conductor other than the neutral one.

Phase leg One of the phase conductors (an ungrounded or "hot" conductor) of a polyphase electrical system.

Phase out A procedure by which the individual phases of a polyphase circuit or system are identified. Someone might "phase out" a three-phase circuit for a motor in order to identify phase A, phase B, and phase C. That person would then know how to connect them to the motor to get the correct phase rotation, causing the motor to rotate in the desired direction.

Phase sequence The order in which the successive members of a periodic wave set reach their positive maximum values: (a) zero phase sequence—no phase shift; (b) plus/minus phase sequence—normal phase shift.

Phase shift The absolute magnitude of the difference between two phase angles.

Photocell A device in which the current-voltage characteristic is a function of incident radiation (light).

Photoelectric control A control sensitive to incident light.

Photoelectricity A physical action wherein an electrical flow is generated by light waves.

Photon An elementary quantity (quantum) of radiant energy.

Pilot lamp A lamp that indicates the condition of an associated circuit.

Pilot wire An auxiliary insulated conductor in a power cable used for control or data.

Plating Forming an adherent layer of metal on an object.

Plug A male connector for insertion into an outlet or jack.

Polarity (1) Distinguishing one conductor or terminal from another. (2) Identifying how devices are to be connected, such as plus (+) or minus (−) signs.

Polarization Index Ratio of insulation resistance measured after 10 minutes to the measure at 1 minute with voltage continuously applied.

Pole (1) That portion of a device associated exclusively with one electrically separated conducting path of the main circuit or device. (2) A supporting circular column.

Polyphase circuits Circuits running on ac and having two or more interrelated voltages, usually of equal amplitudes, phase differences, and periods, etc. If a neutral conductor exists, the voltages referenced to the neutral conductor are equal in amplitude and phase. The most common version is that of three-phase, equal in amplitude with phases 120° apart.

Portable Designed to be movable from one place to another, not necessarily while in operation.

Positive Connected to the positive terminal of a power supply.

Potential The difference in voltage between two points of a circuit. Frequently, one is assumed to be ground (zero potential).

Potential energy Energy of a body or system with respect to the position of the body or the arrangement of the particles of the system.

Potentiometer An instrument for measuring an unknown voltage or potential difference by balancing it, wholly or in part, by a known potential difference produced by the flow of known currents in a network of circuits of known electrical constants.

Power (1) Work per unit of time. (2) The time rate of transferring energy. As an adjective, the word "power" is descriptive of the energy used to perform useful work: pound-feet per second, watts.

Power, active In a three-phase symmetrical circuit, $p = 3\ VI \cos \theta$; in a one-phase, two-wire circuit, $p = VI \cos \theta$.

Power, apparent The product of rms volts times rms amperes.

Power element Sensitive element of a temperature-operated control.

Power factor Correction coefficient for ac power necessary because of changing current and voltage values.

Power loss (cable) Loss due to internal cable impedance, mainly I^2R: the loss causes heating.

Pressure motor control A device that opens and closes an electrical circuit as pressures change.

Primary Normally referring to the part of a device or equipment connected to the power supply circuit.

Primary control Device that directly controls operation of a heating system.

Printed circuit A board having interconnecting wiring printed on its surface and designed for mounting of electronic components.

Process Path of succession of states through which a system passes.

Program, computer The ordered listing of a sequence of events designed to direct the computer to accomplish a task.

Protector, circuit An electrical device that will open an electrical circuit if excessive electrical conditions occur.

Proton The hydrogen atom nucleus; it is electrically positive.

Prototype The first full-size working model.

Proximity effect The distortion of current density due to magnetic fields; increased by conductor diameter, close spacing, frequency, and magnetic materials such as steel conduit or beams.

Pull box A sheet-metal box-like enclosure used in conduit runs, either single conduits or multiple conduits, to facilitate pulling in of cables from point to point in long runs or to provide installation of conduit support bushings needed to support the weight of long riser cables or to provide for turns in multiple-conduit runs.

Pyrometer Thermometer that measures the radiation from a heated body.

Raceway Any channel designed expressly for holding wire, cables, or bars and used solely for that purpose.

Rack (cable) A device to support cables.

Radar A radio detecting and ranging system.

Radiant energy Energy traveling in the form of electromagnetic waves.

Radiant heating Heating system in which warm or hot surfaces are used to radiate heat into the space to be conditioned.

Radiation The process of emitting radiant energy in the form of waves or particles.

Radiation, blackbody Energy given off by an ideal radiating surface at any temperature.

Radiation, nuclear The release of particles and rays during disintegration or decay of an atom's nucleus. These rays—alpha particles, beta particles, and gamma rays—cause ionization.

Radius, bending The radii around which cables are pulled.

Rated. Indicating the limits of operating characteristics for application under specified conditions.

Reactance (1) The imaginary part of impedance. (2) The opposition to ac due to capacitance (Xc) and inductance (XL).

Reactor A device to introduce capacitive or inductive reactance into a circuit.

Receptacle A contact device installed at an outlet for the connection of an attachment plug and flexible cord to supply portable equipment.

Recorder A device that makes a permanent record, usually visual, of varying signals.

Rectifiers Devices used to change alternating current to unidirectional current.

Rectify To change from ac to dc.

Red-leg (1) The phase conductor of a three-phase, four-wire, delta-connected system that is not connected to the single-phase power supply. (2) The conductor with the highest voltage above ground, which must be identified (as per NEC) and is commonly painted red to provide such identification.

Relay A device designed to change a circuit abruptly because of a specified control input.

Relay, overcurrent A relay designed to open a circuit when current in excess of a particular setting flows through the sensor.

Remote-control circuits The control of a circuit through relays and other means.

Resistance The opposition in a conductor to current; the real part of impedance.

Resistor A device whose primary purpose is to introduce resistance.

Resonance In a circuit containing both inductance and capacitance, a condition in which the inductive reactance is equal to and cancels out the capacitance reactance.

Rheostat A variable resistor that can be varied while energized; normally one used in a power circuit.

ROM Read only memory.

Romex General Cable's trade name for type NM cable; but used generically by electrical workers to refer to any nonmetallic sheathed cable.

Roughing in The first stage of an electrical installation, when the raceway, cable, wires, boxes, and other equipment are installed; electrical work that must be done before any finishing or cover-up phases of building construction can be undertaken.

Self-inductance Magnetic field induced in the conductor carrying the current.

Semiconductor A material that has electrical properties of current flow between a conductor and an insulator.

Sensor A material or device that goes through a physical change or an electronic characteristic change as conditions change.

Separable insulated connector An insulated device to facilitate power cable connections and separations.

Service cable The service conductors made up in the form of a cable.

Service conductors The supply conductors that extend from the street main or transformers to the service equipment of the premises being supplied.

Service drop Run of cables from the power company's aerial power lines to the point of connection on a customer's premises.

Service entrance The point at which power is supplied to a building, including the equipment used for this purpose (service main switch or panel or switchboard, metering devices, overcurrent protective devices, conductors for connecting to the power company's conductors, and raceways for such conductors).

Service equipment The necessary equipment, usually consisting of a circuit breaker or switch and fuses and their accessories, located near the point of entrance of supply conductors to a building and intended to constitute the main control and cutoff means for the supply to the building.

Service lateral The underground service conductors between the street main, including any risers at a pole or other structure or from transformers, and the first point of connection to the service-entrance conductors in a terminal box, meter, or other enclosure with adequate space, inside or outside the building wall. Where there is no terminal box, meter, or other enclosure with adequate space, the point of connection is the entrance point of the service conductors into the building.

Service raceway The rigid metal conduit, electrical metallic tubing, or other raceway that encloses the service-entrance conductors.

Sheath A metallic close-fitting protective covering.

Shield The conducting barrier against electromagnetic fields.

Shield, braid A shield of interwoven small wires.

Shield, insulation An electrically conducting layer to provide a smooth surface in intimate contact with the insulation outer surface; used to eliminate electrostatic charges external to the shield and to provide a fixed known path to ground.

Shield, tape The insulation shielding system whose current-carrying component is thin metallic tapes, now normally used in conjunction with a conducting layer of tapes or extruded polymer.

Short-circuit An often unintended low-resistance path through which current flows around, rather than through, a component or circuit.

Shunt A device having appreciable resistance or impedance connected in parallel across other devices or another apparatus to divert some of the current. Appreciable voltage exists across the shunt and appreciable current may exist in it.

Signal A detectable physical quantity or impulse (such as a voltage, current, or magnetic field strength) by which messages or information can be transmitted.

Signal circuit Any electrical circuit supplying energy to an appliance that gives a recognizable signal.

Single-phase motor Electric motor that operates on single-phase alternating current.

Single-phasing The abnormal operation of a three-phase machine when its supply is changed by accidental opening of one conductor.

Solenoid Electric conductor wound as a helix with a small pitch; coil.

Solidly grounded No intentional impedance in the grounding circuit.

Solid state A device, circuit, or system which does not depend on the physical movement of solids, liquids, gases, or plasma.

SP Single pole.

Specs Abbreviation for the word "specifications," which is the written precise description of the scope and details of an electrical installation and the equipment to be used in the system.

Starter (1) An electric controller for accelerating a motor from rest to normal speed and for stopping the motor. (2) A device used to start an electric discharge lamp.

Starting relay An electrical device that connects and/or disconnects the starting winding of an electric motor.

Starting winding Winding in an electric motor used only during the brief period when the motor is starting.

Static Interference caused by electrical disturbances in the atmosphere.

Stator The portion of a rotating machine that includes and supports the stationary active parts.

Steady state When a characteristic exhibits only negligible change over a long period of time.

Strand A group of wires, usually twisted or braided.

Supervised circuit A closed circuit having a current-responsive device to indicate a break or ground.

Surge (1) A sudden increase in voltage and current. (2) Transient condition.

Switch A device for opening and closing or for changing the connection of a circuit.

Switch, ac general-use snap A general-use snap switch suitable only for use on alternating-current circuits and for controlling resistive and inductive loads (including electric discharge lamps) not exceeding the ampere rating at the voltage involved.

Switchboard A large single panel, frame, or assembly of panels having switches, overcurrent and other protective devices, buses, and usually instruments mounted on the face or back or both. Switchboards are generally accessible from the rear and from the front and are not intended to be installed in cabinets.

Switch, general-use A switch intended for use in general distribution and branch circuits. It is rated in amperes and is capable of interrupting its rated voltage.

Switch, general-use snap A type of general-use switch so constructed that it can be installed in flush device boxes or on outlet covers or otherwise used in conjunction with wiring systems recognized by the National Electrical Code.

Switch, isolating A switch intended for isolating an electrical circuit from the source of power. It has no interrupting rating and is intended to be operated only after the circuit has been opened by some other means.

Switch, knife A switch in which the circuit is closed by a moving blade engaging contact clips.

Switch-leg The part of a circuit that runs from a lighting outlet box where a luminaire or lampholder is installed down to an outlet box which contains the wall switch that turns the light or other load on or off; it is a control leg of the branch circuit.

Switch, motor-circuit A switch, rated in horsepower, capable of interrupting the maximum operating overload current of a motor having the same horsepower rating as the switch at the rated voltage.

Synchronous machine A machine in which the average speed of normal operation is exactly proportional to the frequency of the system to which it is connected.

Synchronous speed The speed of rotation of the magnetic flux produced by linking the primary winding.

Synchrotron A device for accelerating charged particles to high energies in a vacuum; the particles are guided by a changing magnetic field while they are accelerated in a closed path.

System A region of space or quantity of matter undergoing study.

Tachometer An instrument for measuring revolutions per minute.

Tap (1) A splice connection of a wire to another wire (such as a feeder conductor in an auxiliary gutter) where the smaller conductor runs a short distance (usually only a few feet, but can be as much as 25 ft) to supply a panelboard or motor controller or switch. Also called a "tap-off," indicating that energy is being taken from one circuit or piece of equipment to supply another circuit or load. (2) A tool that cuts or machines threads in the side of a round hole.

Telegraphy Telecommunication by the use of a signal code.

Telemetering Measurement with the aid of intermediate means that permits interpretation at a distance from the primary detector.

Telephone The transmission and reception of sound by electronics.

Thermal cutout An overcurrent protective device containing a heater element in addition to and affecting a renewable fusible member which opens the circuit. It is not designed to interrupt short-circuit currents.

Thermally protected (as applied to motors) Refers to the words "thermally protected" appearing on the nameplate of a motor or motor-compressor and means that the motor is provided with a thermal protector.

Thermal protector (as applied to motors) A protective device that is assembled as an integral part of a motor or motor compressor and that, when properly applied, protects the motor against dangerous overheating due to overload and failure to start.

Three-phase system A three-phase, alternating-current system containing three individual circuits or phases. Each phase is timed so that the current alternations of the first phase are one-third of a cycle (120°) ahead of the second and two-thirds of a cycle (240°) ahead of the third.

Transformer A device used to transfer energy from one circuit to another. It is composed of two or more coils linked by magnetic lines of force.

Trusses Framed structural pieces consisting of triangles in a single plane for supporting loads over spans.

Utilization equipment Equipment that utilizes electric energy for mechanical, chemical, heating, lighting, or other similar useful purposes.

Ventilated Provided with a means to permit enough circulation of air to remove an excess of heat fumes or vapors.

Volt The practical unit of voltage of electromotive force. One volt sends a current of one ampere through a resistance of one ohm.

Voltage Voltage is the force, pressure, or electromotive force (emf) which causes electric current to flow in an electric circuit. Its unit of measurement is the volt, which represents the amount of electrical pressure that causes current to flow at the rate of one ampere through a resistance of one ohm. Voltage in an electric circuit may be considered as being similar to water pressure in a pipe or water system.

Voltage drop The voltage drop in an electric circuit is the difference between the voltage at the power source and the voltage at the point at which electricity is to be used. The voltage drop, or loss, is created by the resistance of the connecting conductors.

Voltage-to-ground In grounded circuits the voltage between the given conductor and that point or conductor of the circuit which is grounded; in ungrounded circuits, the greatest voltage between the given conductor and any other conductor of the circuit.

Watertight So constructed that moisture will not enter the enclosing case or housing.

Watt The unit of measurement of electrical power or rate of work; 756 watts is equivalent to 1 horsepower. The watt represents the rate at which power is expended when a pressure of one volt causes current to flow at the rate of one ampere. In a dc circuit or in an ac circuit at unity (100 percent) power factor, the number of watts equals the pressure (in volts) multiplied by the current (in amperes).

Weatherproof So constructed or protected that exposure to the weather will not interfere with successful operation.

Web Central portion of an I beam.

INDEX

NOTE: Page numbers in *italics* indicate illustrations.

About the Author

John E. Traister has been involved in the electrical construction industry (including the design and installation of security and fire-alarm systems) for more than 20 years. He operated his own electrical contracting business for a number of years, and he has also worked as an electrical designer, estimator, and installer. In addition to writing over 100 articles for trade and professional journals, Mr. Traister is the author of more than 100 books, including *Electrical Design for Building Construction* (McGraw-Hill).